Proofs and Logical Arguments Supporting the Foundational Laws of Physics

For scientists, students, and curious laypersons, this compilation, *Proofs and Logical Arguments Supporting the Foundational Laws of Physics: A Handy Guide for Students and Scientists* examines the most important laws and relationships taught in science courses, attaching a short and accessible proof or logical argument for each assertion.

Every thoughtful person should seek to understand why we think we know what we say we know about the natural world. Otherwise, we may as well surrender ourselves to a world ruled by magic. In 136 essays, readers are provided with proofs and logical arguments supporting the laws and relationships that serve as the foundation of our rational understanding of reality. Among the essays included in this book, we will find proofs of Pauli's exclusion principle, Heisenberg's uncertainty principle, the principles of special relativity, the Schrodinger wave equation, Noether's theorem, and many of the laws of physics and chemistry that no scientist should accept on blind faith alone.

Laypersons will find that the ideas discussed in this volume are always thought-provoking and sometimes inspiring. For university undergraduates, the book will serve as an introduction to the core sciences. Graduate students may find this book to be a handy cross-disciplinary reference that explains how the tools of their own selected discipline have emerged from fundamental principles that unify all the sciences.

Jules J. Berman received two baccalaureate degrees from MIT (from the Department of Mathematics, and from the Department of Earth and Planetary Sciences). He holds a PhD from Temple University, and an MD, from the University of Miami. His postdoctoral studies were completed at the US National Institutes of Health, and his residency was completed at the George Washington University Medical Center in Washington, DC. Dr. Berman served as Chief of Anatomic Pathology, Surgical Pathology, and Cytopathology at the Veterans administration Medical Center in Baltimore, Maryland, where he also held joint appointments at the University of Maryland Medical Center and at the Johns Hopkins Medical Institutions. In 1998, he transferred back to the US National Institutes of Health, as a Medical Officer, and as the Program Director for Pathology Informatics in the Cancer Diagnosis Program at the National Cancer Institute. Dr. Berman is a past president of the Association for Pathology Informatics, and is the 2011 recipient of the Association's Lifetime Achievement Award. He has first-authored more than 100 journal articles and has written more than 20 single-author science books.

Proofs and Logical Arguments Supporting the Foundational Laws of Physics

A Handy Guide for Students and Scientists

Jules J. Berman

CRC Press is an imprint of the
Taylor & Francis Group, an **informa** business

Designed cover image: Shutterstock_2052374897

First edition published 2025
by CRC Press
2385 NW Executive Center Drive, Suite 320, Boca Raton FL 33431

and by CRC Press
4 Park Square, Milton Park, Abingdon, Oxon, OX14 4RN

CRC Press is an imprint of Taylor & Francis Group, LLC

ISBN: 978-1-032-85067-2 (hbk)
ISBN: 978-1-032-85068-9 (pbk)
ISBN: 978-1-003-51637-8 (ebk)

DOI: 10.1201/9781003516378

Typeset in URWPalladioL-Roma font
by KnowledgeWorks Global Ltd.

Dedication

For Noah, who may one day find this book and read it.

Contents

List of Figures

Author Biography

Jules J. Berman received two baccalaureate degrees from MIT (from the Department of Mathematics, and from the Department of Earth and Planetary Sciences). He holds a PhD from Temple University, and an MD, from the University of Miami. His postdoctoral studies were completed at the US National Institutes of Health and his residency was completed at the George Washington University Medical Center in Washington, DC. Dr. Berman served as Chief of Anatomic Pathology, Surgical Pathology, and Cytopathology at the Veterans Administration Medical Center in Baltimore, Maryland, where he also held joint appointments at the University of Maryland Medical Center and at the Johns Hopkins Medical Institutions. In 1998, he transferred back to the US National Institutes of Health, as a Medical Officer, and as the Program Director for Pathology Informatics in the Cancer Diagnosis Program at the National Cancer Institute. Dr. Berman is a past president of the Association for Pathology Informatics and is the 2011 recipient of the Association's Lifetime Achievement Award. He has first-authored more than 100 journal articles and has written more than 20 single-author science books.

Books by Jules J. Berman

Preface

"Before I speak, I have
something important to say".
—Groucho Marx

Much of what we think we know about the operating system of the universe comes to us as "received wisdom"—settled knowledge whose truth cannot be questioned. For example, we are informed that all objects fall at the same rate, regardless of their size or mass. We are told that moving matter must travel at speeds less than the speed of light. We are assured that our observable universe began in a single instant, about 13.8 billion years ago, in the form of a Big Bang. For most of us, these assertions are pleasant factoids; items that we accept as truth, but which play no role in our daily lives. Our ability to get passing grades in high school and college depends very much on our blind acceptance of the information recited by our teachers and written in our textbooks. Too few of us feel compelled to find the logical arguments that account for the fundamental principles of nature.

It just so happens that the scientific world has reached a point where we can posit reasonable proof for the underlying principles of physics. Such proofs, for the most part, are based upon logical arguments applied to simple observations. No experimental data is required to prove that objects of any weight and composition will fall at the same speed, or that objects must move more slowly than the speed of light in a vacuum. No measurements are necessary to explain the properties of the fundamental constants of nature. The famous experiments that establish the textbook rules of physics are, in many cases, performed solely for the purpose of validating proofs based entirely on theory.

What is "proof" in the physical sciences, and how does it differ from proof in mathematics?[1] A physical proof is a logical argument, stemming from some condition (often an initial condition characteristic of the system being studied) that leads to an assertion or law. The proof is invalid if our

[1]The physical sciences loosely encompass quantum physics, mechanics, thermodynamics, electromagnetism, wave theory, electricity, physical chemistry, and cosmology. These are the disciplines that emerge from the most elementary properties of spacetime. All of these fields rely heavily on mathematics, which traditionally occupies its own intellectual territory, outside the physical sciences.

understanding of the initial conditions are in error, or if our logic is in error, or if the resulting law is not generalizable (i.e., not always true).[2] In mathematics, proofs are based upon some initial postulates that are accepted as an abstraction of fact. That is to say, that the truth of the initial postulates of mathematics need not correspond to any physical reality. The postulates are what we say they are, and are never true or false. From the postulates, we build logical arguments and develop theorems that are logically proven.

We can see that all physical proofs occupy a state of scientific sin. We cannot fully trust our proofs, because they all depend upon a perfect understanding of the fundamental conditions of reality. Our concepts of reality are interpretations produced by our limited and fallible minds and cannot be accepted as absolute truths. In some ways, having proof that we cannot fully believe is a good thing. It allows us to continuously question what we think we know with certainty, and these questions often lead to new ways of understanding our world. To emphasize this point, let's imagine, for a moment, that this book provides an immaculate rendition of the fundamental properties of the physical world, along with a perfect set of proofs that establish all of the laws of the universe. Because the complexities of the natural world emerge from interactions among fundamental forces and particles, we could claim that everything we can ever know about ourselves and our world is accounted for in the pages of this book. If that were the case, we could dismiss all of the sciences as predictable derivatives of the proofs contained herein. With all of our questions answered, and with nothing new on the horizon, we might as well suspend the pursuit of science and take up stamp collecting. Luckily, today's fundamental proofs are subject to change without notice. Science is an eternal mystery.

Given the aforementioned limitations on "proof", who might benefit from reading this imperfect book? Every thoughtful person should seek to understand why we think we know what we say we know about the natural world. Otherwise, we may as well surrender ourselves to a world ruled by magic. This book reminds us that the scientific beliefs we hold are based upon reasonable arguments that can be analyzed and revisited whenever we please. The laws and relationships in this volume serve as the foundation of our rational understanding of reality and have permitted humans to achieve the technological wonders that distinguish Homo sapiens from all other species. Laypersons may find that the ideas discussed in this volume are always thought-provoking and sometimes inspiring. University

[2] Throughout this book, we take the liberty of referring to the steps leading from an observation to a conclusion (usually in the form of a law an equation, or a principle) as "proof," in the restrictive sense that the conclusion can be logically defended. The proofs are intended to provide readers with a scientific justification for many of the most important concepts in the physical sciences.

undergraduates will find this book to be a good introduction to all of the core sciences. Graduate students may find this book to be a handy cross-disciplinary reference that explains how the tools of their own selected discipline have emerged from fundamental principles that unify all the sciences.

1

Particles and Atoms

> *"What really motivates elementary particle physicists is a sense of how the world is ordered - it is, they believe, a world governed by simple universal principles that we are capable of discovering".*
>
> —Steven Weinberg

This chapter describes the properties of fundamental particles. In addition, the chapter provides logical arguments explaining why these properties must exist as they do. In this chapter, the proof is offered to show that fundamental particles are not composed of matter, have no size, and never contain other fundamental particles. Despite these deficiencies, such particles account for all of the observable matter in the universe. Among the many proofs offered in this chapter, readers will find proof that all fundamental particles of a kind are identical to one another, a short proof of the Pauli exclusion principle (i.e., no two identical particles, having the same quantum state, may occupy the same location), proof that fundamental particles of a kind can not actually collide with one another, and a proof that there is an upper limit to the number of different kinds of elements in the universe.

DOI: 10.1201/9781003516378-1

1.1 Proving fundamental particles do not contain matter

"What is mind? Doesn't matter.
What is matter? Never mind".
—Nihilistic folk wisdom

The proof that fundamental particles do not contain matter is simple:[1]

1. By definition, a fundamental particle is one which cannot be divided into smaller particles.
2. If a fundamental particle contained matter, then it could be split into the matter composing itself.

   Therefore,
3. The particle would not be fundamental, contradicting our original premise that the fundamental particle contained matter.

   Therefore,
4. The fundamental particle cannot contain matter.

We can find a second proof that fundamental particles do not contain matter.

1. Fundamental particles of a particular type are identical (see section titled "Proving the identicality of fundamental particles of a kind")
2. If the fundamental particles were made from matter, then particles of a type would not be exactly identical to one another.

   We might expect to have one particle containing just a bit more or less "stuff" than the next particle. Or a particle might lose some of its "stuff" over time. Or relativistic effects would cause the particle to change its mass as it moved.[2]

   Therefore,
3. Fundamental particles of any type, being identical to one another, cannot contain matter.

---

[1]The notion that fundamental particles do not contain matter was suggested as early as 1759 by Roger Boscovich (1711-1787) in his manuscript, *Philosophiae naturalis Theoria redacta ad unicam Legem Virium*. Boscovich also reasoned, quite correctly, that it would be impossible for any fundamental particle to have size.

[2]We will revisit the point in the section titled "Proving the identicality of fundamental particles of a kind".

If fundamental particles contain no matter, then what are they? We will learn in subsequent sections that fundamental particles are best conceived as waves that have properties characteristic of their particular fields.[3] Waves are propagating disturbances in a field. There is no "stuff" in a wave. Consistent with the wave notion of particles, we note that fundamental particles emit waves (e.g., light is emitted by electrons moving from one atomic shell to another shell having a lower energy).[4] Alternately, waves can emit particles.[5]

As an aside, we can infer that if there is no matter within fundamental particles, then the fundamental particles have no size. This being the case, we are confronted with a dilemma. All non-fundamental particles are composed of fundamental particles. For example, atoms are made of protons and neutrons (both non-fundamental) and electrons (fundamental), accounting for nearly all of the matter observable in the universe. Protons and neutrons, in turn, are composed of quarks and gluons (both fundamental particles).[6] Because matter is composed of fundamental particles that have no size, then we would expect that matter itself should have no size! Where did we go wrong in our thinking?

Although matter is composed of point particles (sizeless particles), the point particles operate under forces that attract or repel other particles, and which account for the boundaries between different types of particles. The Pauli exclusion principle prohibits identical massive particles from occupying the same spacetime location. Thus matter has size. Particles that move at the speed of light (e.g., photons) have no size **and** are not subject to the exclusionary principles that govern particles that travel at less than the speed of light (i.e., massless particles have the property of superpositioning).

---

[3]The three fields described by the Standard Model of quantum physics are the electromagnetic field, the strong field, and the weak field. Each field has interacting forces carried by field-specific particles. For example, the W and Z particles carry the weak force, and gluons carry the strong force. A fourth field, gravitation, is not addressed in the Standard Model of quantum physics.

[4]Strictly speaking, particles do not "emit" anything. The light leaving the particle is just part of the description of the wave particle.

[5]See Burke DL, Field RC, Horton-Smith G, Spencer JE, Walz D, Berridge SC, et al. Positron production in multiphoton light-by-light scattering. 79:1626, 1997.

[6]Quarks account for greater than 99% of the matter in the observable universe.

1.2 Proving fundamental particles contain no other fundamental particles

If the fundamental particles are just collections of the properties of their respective fields, and are not composed of "stuff", then how is it possible for one fundamental particle to decay into several different particles? [7] For example, we know that one tau particle, τ, can decay into several fundamental particles (electrons, neutrinos of different types, quarks, pions). Particle physicists seem to be the only humans capable of keeping track of all the options. If fundamental particles are composed of other particles, then it would seem that such particles cannot be composed of "nothing" (despite our prior proof to the contrary).

Well, no, when a particle decays, it does not decay into its constituent parts (it has no constituent parts). Fundamental particles decay into other particles in accordance with its properties (remember, fundamental particles have field-specific properties).

Let's look at the particular case of τ (tau) particle decay.

1. τ particles may decay in numerous ways, to produce many different combinations of decay products. Examples of various options for τ particle decay are:
    - $\tau \rightarrow$ a charged pion, a neutral pion, and a τ neutrino
    - $\tau \rightarrow$ a charged pion and a τ neutrino
    - $\tau \rightarrow$ a charged pion, two neutral pions, and a τ neutrino
    - $\tau \rightarrow$ three charged pions, and a τ neutrino
    - $\tau \rightarrow$ three charged pions, a neutral pion, and a τ neutrino
    - $\tau \rightarrow$ three neutral pions, a charged pion, and a τ neutrino
    - $\tau \rightarrow$ a τ neutrino, an electron, and an electron anti-neutrino
    - $\tau \rightarrow$ a τ neutrino, a muon, and a muon anti-neutrino
2. If τ particles (all of which are identical to one another) contained sub-particles, then we would expect to find only one possible set of τ decay products.

   Otherwise, one τ particle would be non-identical to another τ particle following some alternate decay option.

[7] When we use the term "decay" here, we do so to indicate an interaction of the particle with its environment; not some internal process that leads to the particle's self-annihilation.

3. Because there are many options for τ decay, including non-overlapping sets of particle products, we infer that τ particle could not have **contained** any particular set of sub-particles.
4. Therefore, we conclude that the decay products of τ particles are not contained within the τ particle.

We cannot open up a τ particle and expect a neutrino and a pion to pop out! As it happens, particle formation is a property of the particle's field. When one particle decays, and other particles emerge, we can surmise that field interactions are responsible for the emergence of the new particles, contributed by several fields.

Aside from particles yielding other particles, we can also see non-particles yielding particles. When an electron and an anti-electron (i.e., positron) encounter one another, they mutually annihilate, yielding energy in the form of light. In a reversal of roles, high-energy photons on a collision course may yield electron-positron pairs.[8] Thus, particles of matter can be created from an energetic electromagnetic field.

[8]Sugimoto discusses the creation of positrons from light via the Breit-Wheeler effect. See Sugimoto K, He Y, Iwata N, Yeh I-L, Tangtartharakul K, Arefiev A, Sentoku Y. Positron generation and acceleration in a self-organized photon collider enabled by an ultraintense laser pulse. Phy Rev Lett 131:065102, 2023.

1.3 Proving the identicality of fundamental particles of a kind

"I always wanted to be somebody,
but now I realize I should have been more specific".
—Lily Tomlin

Physicists have proposed several different ways to describe the physical properties that define a fundamental particle. Let's look at a few.

- A quanta of a field
- The minimal observable excitation of a field
- A wave packet

  A wave packet is a summation of waves whose frequencies are close to one another, which, when superpositioned, yields something akin to a discrete pulse that has the features of a particle. We will be discussing wave packets in more detail in the sections titled "Proving that fundamental particles can be interpreted as wave packets" and "Proving Fourier transforms of Gaussians are Gaussians".
- A probability distribution for the properties of fields that can be conveniently modeled by a wave packet
- A collapsed wave function (the result of the wave model at the moment when it fails)
- Whatever is detected in a particle detector

  Fundamental particles, having no size, cannot be seen. The best we can ever do is to see the point-like effect of the particle in a detector designed to register some observable property of the moving particle. The localized property registered in the detector is, for all practical purposes, equivalent to the particle.
- A wave that has lost its coherence with other waves
- A conserved property (such as a charge, in the case of the electron) near a point, having a non-stationary distribution
- A conserved property in a fixed ratio to its mass (in the case of particles with mass)
- A wave having quantized rotational motion (spin)

  Fundamental particles, having no size, cannot spin, but there is a quantized angular momentum that must be assigned to each type of particle, and this angular or rotational momentum is a property of waves (such as electromagnetic waves).

- An observable capable of entanglement

  The law of conservation of angular momentum dictates that the spin of one particle must be the opposite of the spin of its simultaneously created pair. In this case, the two fundamental particles are said to be entangled. When the spin of either of the particles is determined, the spin of the other particles must be equal but opposite.

- A manifestation of the vibrational state of a quantum string

We needn't delve into the specifics of each definition. Suffice it to remark that all of the seemingly disparate definitions are, in fact, logical and necessary consequences of one another. Perhaps the definition that is easiest to understand is that a particle is the smallest vibration of a field. This definition implies that fields have a "smallest" vibration beneath which nothing is observable and nothing is capable of carrying sufficient energy to have any effect. Put simply, when a vibration is smaller than the smallest, it ceases to exist. We can now develop an intuitive proof that fundamental particles of a kind are identical to one another.

1. We start with the current definition of a particle as the smallest possible vibration of a field (i.e., the quantum of the field).
2. But there can only be one "smallest" vibration.[9]
3. Therefore, all particles of a field have the same vibration.
4. Because a particle has no size, it is characterized by its movement (as vibration) and its associated energy (determined by the vibration).
5. Therefore, all particles of a field must be identical.

Is there any experimental evidence leading to the same conclusion? Yes, indeed. The experiment-based argument for the identicality of fundamental particles hinges on the second law of thermodynamics, which holds that the entropy of isolated systems left to spontaneous evolution always achieves a state of thermodynamic equilibrium for which the entropy is maximal (i.e., maximal particle randomness).[10]

[9]See section titled "Proving that there is a minimal particle frequency below which no energy is observable".

[10]A simple way of thinking of entropy is as a measure of the information required to fully describe the components of a system. Increasing the number of components of a subsystem would increase its entropy, as would increasing the degrees of freedom by which those components may freely move.

A perplexing phenomenon occurs when we compare the change of entropy that occurs when mixing volumes of gases; the so-called mixing paradox. When there are two different types of gases, held under the same conditions (volume, pressure, temperature, and number of atoms), the total entropy of the system, when the two different gases are mixed, will increase. If we mix two gases of the same type, the entropy of the mixture does not increase. The mixing paradox is an apparent paradox insofar as we would expect any mixture of gases to increase the entropy of the system, in accordance with the second law of thermodynamics.[11]

The mixing paradox is an empirical fact with a very simple physical explanation.

1. Whenever we mix two volumes of different gases, the entropy of the system increases as the molecules of gas increase in volume, number, and variety, and the total information required to describe the mixed system (i.e., the entropy of the mixed system) is greater than the information required to describe the distribution of molecules of either single gas.

2. When we mix two volumes of the same gas, the resulting mixture consists of interchangeable molecules, and the information required to describe the distribution of the mixture remains unchanged.

   But,

3. The presumed interchangeability of molecules of the same kind assumes that such molecules are **absolutely** identical.

   If, for example, we have a bottle filled with Argon atoms, the interchangeability of one Argon atom with another requires that each Argon atom must be absolutely identical to one another. Otherwise, despite all the atoms having the same name, their differences would increase the entropy of the system when mixed together.

   Therefore,

4. All of the molecules of any specific gas must be identical to one another under the mixing paradox.

   But,

5. Every molecule of a gas is composed of atoms, and every atom is composed of fixed numbers of several fundamental particles.

[11]The mixing paradox was introduced by Josiah Gibbs for the Connecticut Academy of Sciences, 1875-1878, as "On the equilibrium of heterogeneous substances" and republished in "Selected Papers on Thermodynamics and Statistical Physics", Cambridge University Press, London, 1951. The mixing paradox is also known as the Gibbs paradox.

Therefore,

6. **The mixing paradox tells us that every fundamental particle of a particular kind is absolutely identical to every fundamental particle of the same kind.**

1.4 Proving the Pauli exclusion principle

"In an atom, there cannot be two or more equivalent electrons, for which in strong fields the values of all four quantum numbers coincide".
—Wolfgang Pauli, 1925

Because fundamental particles have no size, it would seem that we could put all of the fundamental particles in the universe in one spot, without any real crowding. At the very least, we should be able to push any two objects together, so that they share the same volume. Casual observation informs us that although the particles that compose our universe have no size, they manage to keep their distance from one another, under most circumstances.[12] The Pauli exclusion principle asserts that two identical particles, having the same quantum states, cannot both occupy the same location. For this reason, solid objects cannot be superpositioned into one another.

It would seem a proof of the Pauli exclusion principle would require us to master the field of quantum physics. Luckily for us, all we need for our proof is some understanding of symmetric and anti-symmetric functions.

A function is symmetric if its value stays the same when the variable that it acts upon changes its sign. A function is anti-symmetric when the value of a function changes its sign when its variable changes its sign.[13]

$$f(x) = f(-x) \qquad \text{Symmetric function}$$

$$f(x) = -f(-x) \qquad \text{Anti-symmetric function}$$

The function that describes our knowledge of a quantum system is its wave equation, Ψ, and the absolute value of the square of a wave function tells us the wave probability density. If we were to sum the wave probability density over all possible quantum states, the probability must add up to 1, as in $\int_{-\infty}^{\infty} |\psi(x)|^2 = 1$, and this total probability must include all allowable states, whether they are symmetric or antisymmetric.[14]

$$|\psi(x)|^2 \qquad \text{Wave probability density}$$

[12]The exception being Bose-Einstein conjugates. For further information on this exotic topic, see: Anderson MH, Ensher JR, Matthews MR, Wieman CE, Cornell EA. Observation of Bose-Einstein condensation in a dilute atomic vapor. Science 269:198-201, 1995.

[13]Mathematicians manage to derive many useful theorems that apply that distinguish the properties of symmetric and anti-symmetric functions (also known as even and odd functions). Several of these theorems determine how the Fourier transform, named for Joseph Fourier, is calculated for symmetric and anti-symmetric functions, and these theorems are frequently applied in the realm of particle physics.

[14]We will use this relationship again when we derive the Heisenberg Uncertainty Principle.

The wave probability density for a quantum system containing two identical particles does not change when we switch one particle with another (that is what its identicality means). Here is the equation for the wave probability densities when particle a is switched for an identical particle b.

$$|\psi(x_{a,b})|^2 = |\psi(x_{b,a})|^2$$

This condition holds true under two circumstances: Circumstance 1. When we switch a and b and the wave function does not change,

$$\psi(x_{a,b}) = \psi(x_{b,a})$$

Circumstance 2. When we switch a and b, and the wave function changes its sign.

$$\psi(x)_{a,b} = -\psi(x)_{b,a}$$

Circumstance 1 is the symmetric form of the wave function and circumstance 2 is the anti-symmetric form. Both solutions (symmetric and anti-symmetric) represent the allowable states of the identical particles. Both symmetric and anti-symmetric solutions must exist, and both solutions must be included in the summation of the wave probability densities over all possible locations.[15]

Let's imagine a situation in which we have two identical particles (particle1 and particle2) occupying the same quantum state (i.e., have the same quantum properties), at the same location. In this case, their wave equation will be unchanged when we interchange the particles.

$$\psi(x_{particle1,particle2}) = \psi(x_{particle2,particle1})$$

The value of the wave equation does not change when particles are switched because particle1 and particle2 are identical and indistinguishable in every way, including quantum state and location. This seems well and good until we remind ourselves that the wave probability density must have an anti-symmetric solution (where the sign of ψ changes when particles are interchanged).

Knowing this, let's pursue the following line of thought:

1. We have just shown that the sign of ψ does not change when identical particles occupying the same quantum state at the same location are exchanged.

   Therefore,

[15] For proof that symmetric and anti-symmetric forms of all functions must exist, see the section titled "Proving even and odd decomposition applies to all functions".

2. The interchange of identical particles having the same quantum state and location cannot produce an anti-symmetric solution.

   But,

3. An antisymmetric solution is required.

   Therefore,

4. Our premise, that two identical particles having the same quantum state may co-exist in the same location, is impossible.

   Equivalently,

5. **No two identical particles, having the same quantum state, may occupy the same location.**

This is the essence of the Pauli exclusion principle. By playing a bit with the meaning of identicality, we have proven the principle that accounts for the composition and qualities of our material universe.

1.5 Proving the atomic particle theory

"In chemistry we are often setting out to understand what has actually occurred rather than deliberately contriving to fulfil predictions. We are, so it has been said, telling 'likely stories' rather than hazarding and testing prophecies".
—D.W. Theobald[16]

In the first decade of the 19th century, John Dalton, building on the works of Antoine Lavoisier and Joseph Proust, published an atomic theory of matter. His theory contains several bold assertions:

- All matter is made of atoms, which are indivisible.
- The atoms of any element are identical to one another.
- Compounds are composed of combinations of atoms of elements.
- Chemical reactions are just re-distributions of atoms among reactants (no atoms gained or lost).

Dalton knew nothing about isotopes of atoms and subatomic particles, but let's not quibble. Dalton's atomic theory was an astonishing scientific breakthrough, but it remained an unproven theory until the 20th century. In the interim, some of the greatest advances in the field of physics and chemistry were performed speculatively, on the expectation that atoms were real. What brought scientists to believe in the atomic theory, before it was proven?

- The polygonal shape of repeating crystal units

  In the 19th century, one of the best arguments, short of proof, for the existence of atoms came from observations of crystals. Because all naturally occurring crystals are composed of repeating units of a particular polygonal shape, it was concluded that discrete atoms occupied specific positions with respect to one another.[17] Knowing that crystals are composed of the same elements found in gases and liquids and noncrystallizing solids, it seemed reasonable to infer that if crystals are composed of atoms, then all matter is composed of atoms.

- Stoichiometric chemical reactions

  If matter were composed of material of a non-particulate nature, then there would be no compelling reason for chemical reactions to proceed

[16]Theobald DW. Some Considerations on the Philosophy of Chemistry. Chem Soc Rev 5:203-213, 1976.

[17]Further work, in the 19th century, showed that all crystal shapes were described by repeating chemical units belonging to a finite set of classes, clarifying the strict structural constraints imposed by positioning atoms in crystal lattices.

with specific proportions of constituent materials. For example, there would be no reason to have water molecules composed of exactly two parts Hydrogen and one part Oxygen.

- Einstein's theoretical explanation of Brownian motion (1905) required the existence of discrete atomic particles

Brownian motion is the random short movements of particles suspended in water. Einstein reasoned that any undisturbed particle in water should either fall as sediment or float upwards, or lay in a still suspension. This is not the case. Suspended particles bounce around, to and fro, up and down, in short, random jerky movements known as Brownian motion. Einstein reasoned (in 1905) that Brownian motion could only be caused by constant collisions between water molecules and the suspended particles. He further reasoned that such collisions could only occur if the atomic theory of matter held. Specifically, discrete atoms of water, agitated by heat, bounce constantly against suspended particles, causing them to move in rapid, random movements.

- Jean Perrin's experimental confirmation of Einstein's theory of Brownian motion.

In 1909, Jean Perrin provided experimental proof of the existence of atoms based on the observed effects of Brownian motions[18]

Perrin was able to determine Avogradro's number (the number of molecules in a mole) with great precision, for any of the types of suspended particles or solutions he cared to study. The formula used by Perrin is:

$$D = \frac{RT}{N} \times \frac{1}{6\pi\alpha\xi}$$

Here, D is the coefficient of diffusion, R is the universal gas constant, and T is the absolute temperature, N is Avogadro's number (about 6×10^{23}), ξ is the viscosity of the fluid, and α is the radius of spherules (e.g., adius of a water molecule). Because both the radius of molecules and the number of molecules could be computed from observations of Brownian motions, it seemed clear that atoms must exist.

- X-ray crystallographic analysis (1912) demonstrated the atomic nature of crystals.

In 1912, the atomic theory became the atomic fact when x-ray crystallographers reconstructed the position of atoms in individual crystal molecules.

[18] A detailed discussion is found in Perrin's book, *Les Atomes*, Libraire Felix Alcan, Paris, 1913.

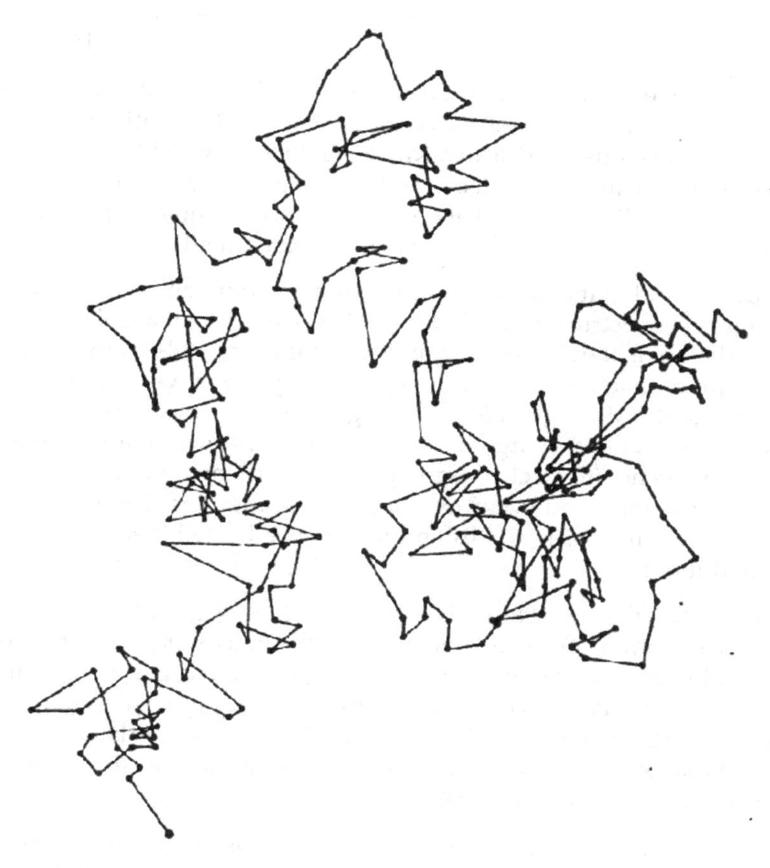

FIGURE 1.1: Jean Perrin measured the perturbations of various suspended solids. The path of particles, studied at 30 second intervals, indicated random motion. Had Perrin reduced the time interval for observation (e.g., 10 seconds, 1 second, 1 millisecond), the picture would look much the same, indicating that molecules bounce off one another, as we would expect from particles. Source, Figure 9 from Jean Perrin's *Les Atomes*, 1913.

1.6 Proving electrons in any given atomic orbital have the same frequency and energy

"In the particular is contained the universal".
James Joyce

In 1860, the German physicist Gustav Robert Kirchhoff demonstrated that black bodies (heated non-reflective objects) emitted a continuous spectrum of radiation. Just one year earlier, Kirchhoff demonstrated that individual elements, when heated, emitted discrete wavelength of radiation that could be seen as spectral lines. Each element had its own unique pattern of spectrographic lines, by which the element could be identified.[19]

Why do we find that black body radiation is random and continuous, while the emission spectrum of an individual element is always discrete? Black body radiation is thermal energy, and thermal energy (also known as heat) is nothing more than kinetic energy, produced by the movement of mass, on a molecular scale. Molecules contain charged particles (electrons and protons), and the movements of charges (in the form of charge accelerations and dipole oscillations) generate electromagnetic waves. The movements are, as far as we know, random, and occur over a range of energies. The resulting spectrum of electromagnetic waves emitted from a heated black body is broad and continuous.

In the case of the discrete emission spectrum of individual elements, the emitted electromagnetic waves are produced when excited orbiting electrons drop to a lower orbit (more precisely, a drop to a quantum state having lower energy). The mechanisms of electromagnetic emissions observed for an individual chemical element is different than that seen with black bodies composed of many different kinds of molecules, each molecule being composed of many different individual elements.

To begin to understand the physics of orbital transitions, we must focus our attention on an improbably influential schoolteacher named Johann Jakob Balmer (1825-1898).[20] Balmer, like many others of his time, was fascinated by the discrete set of emission lines characterizing each element. In about 1880, Vogel and Huggins published the emission frequencies of hydrogen (i.e.,

[19]Spectroscopic data for individual elements is publicly available at the U.S. National Institute of Standards and Technology (NIST). See Kramida, A., Ralchenko, Yu., Reader, J., and NIST ASD Team. NIST Atomic Spectra Database (ver. 5.11), NIST, 2023

[20]Johann Jakob Balmer earned his living as a teacher in a school for girls, working in obscurity, without benefit of laboratory or influential colleagues.

Ion	Observed Wavelength Air (nm)	Ritz Wavelength Air (nm)	Rel. Int.
Rn I	391.720		10
Rn I	394.172		10
Rn I	395.236		10
Rn I	422.606		10
Rn I	430.776		80
Rn I	433.578		7
Rn I	434.960	434.960+	100
Rn I	443.505		40
Rn I	445.925		50
Rn I	450.848		50
Rn I	457.772		50
Rn I	460.938		50
Rn I	472.176		30
Rn I	572.258		6
Rn I	606.192		10
Rn I	620.075		6
Rn I	638.045		6
Rn I	655.749		10
Rn I	660.643		10
Rn I	662.723		15
Rn I	666.960		6
Rn I	670.428		8
Rn I	675.181		20
Rn I	680.679		6
Rn I	683.695		8
Rn I	683.757		8
Rn I	689.116		10
Rn I	699.890		10

FIGURE 1.2: The emission spectrum of Radon (Rn), for the visible wavelengths. Emissions are restricted to specific wavelengths. The most intense emission occurs at 434.960 nm. Source: Public Domain image, courtesy of the U.S. National Institute of Standards and Technology.

the hydrogen spectroscopic emission lines).[21] In 1885, Johann Jakob Balmer, obtaining a copy of those frequencies, developed a formula that precisely expressed frequency in terms of the numeric order of its emission line (i.e., n = 1,2,3,4, and so on).

$$\frac{1}{\lambda} = R_{\mathrm{H}} \left(\frac{1}{2^2} - \frac{1}{n^2} \right) \quad \text{for } n = 3, 4, 5, \ldots$$

Balmer's attempt at data modeling produced one of the strangest equations in the history of science. At the time, there was simply no precedent for expressing the frequency of an electromagnetic wave in terms of an integer rank. Balmer himself admitted that he was just playing around with numbers, and his formula was introduced to the world without the benefit of any theoretical explanation. Nonetheless, he had hit upon an equation that precisely described multiple emission lines, in terms of ascending integers. **Twenty-eight years later, Niels Bohr, in 1913, chanced upon Balmer's formula and used it to explain spectral lines in terms of energy emissions resulting from transitions between discrete electron orbits.** Balmer's amateurish venture into data analysis led, somewhat inadvertently, to the birth of modern quantum physics.[22]

With the information at hand, we are prepared to prove that the electrons occupying any given atomic orbital have the same frequency and energy:

1. Emissions of photons from excited atoms have discrete, characteristic frequencies.
2. The frequencies of the emitted photons result from electrons dropping from higher energy orbitals to lower energy orbitals. The high-energy orbitals are reached when the element is heated in the spectrograph (gaining energy). The lower energy orbitals are reached when the element cools off (and loses energy).
3. Because the emitted photons have discrete, characteristic frequencies, and each specific photon frequency is associated with a specific energy, and the specific energy corresponds exactly to the energy lost by electrons shifting orbits, we can infer that the electrons shifting from one particular orbit to another particular orbit must all have the same energy.

[21]For the original manuscripts, now classics of science, see Vogel HW. Monatsbericht der Königlichen Academie der Wissenschaften zu Berlin, July 10, 1879; and Huggins W. On the Photographic Spectra of Stars. Phil Trans R Soc Lond 171:669-690, 1880.

[22]Bohr went so far as to derive the Rydberg constant, R (empirically determined for Balmer's formula) from theory alone, showing that $R = 2\pi m q^4 \div h^3$ where m is the mass of an electron, q is the charge of an electron, and h is the Planck constant. Doing so removed the Rydberg constant from the list of measurable but inexplicable constants and placed it among the constants logically derived from the physical relationships of constant quantities.

4. Consequently, the emission spectrum of a heated atom displays as discrete bands corresponding to the identical frequencies of the identical photons emitted during the identical orbital transitions executed by the identical electrons in identical atoms.

1.7 Proving there is a minimal particle frequency below which no energy is observable

In the early Bohr model of the atom, charges moved around the nucleus in circular orbits. The orbits were divided into shells that were numbered 1,2,3, . . . , with 1 being the shell closest to the nucleus. Most children are taught the early Bohr model of the atom because it can be easily compared to the orbits of planets around the sun.

Among the many drawbacks of the early Bohr model of the atom is that, according to classical electrodynamic theory, electrons in movement about the nucleus must emit electromagnetic radiation, causing them to lose energy. Orbiting electrons, attracted to the nucleus by charge, must therefore spiral into the nucleus as they lose energy. If the early Bohr model were true, then electrons would be absorbed into protons and atoms would not exist.

A new model of the atom was needed, and in the first three decades of the twentieth century, physicists feverishly developed the quantum model of the atom, and this model has stood the test of time. Let's review those features of the canon that are most relevant to the discussion in this section:

- The ground state orbital of the atom (corresponding to the first shell in the Bohr atom) is stable and represents the absolute minimum energy level that an electron may possess and still exist.
- One quantum is just another way of indicating the minimum energy requirement for the existence of a field particle. For example, one electron is one quantum of charge. One photon is one quantum of the electromagnetic field. One quark is one quantum of the quark field.
- Matter particles are wave packets (sometimes referred to as matter waves).
- The location of matter particles is probabilistic, and is described by the Schrödinger wave equation.
- For atomic electrons in the ground state (i.e., the lowest possible energy state), the position of the electron is anywhere and everywhere in a sphere that includes the nucleus.
- The ground state orbit of the electron is stable, and electrons in the ground state never drop into the nucleus. We can actually prove this point.

    1. We have previously shown that the energy released when an electron shifts from a higher orbital to a lower orbital is

quantized (see section titled "Proving that the energy levels of electron orbitals are discrete and quantized").

2. If an electron were to drop from the ground state into the nucleus, then it must release a quantized amount of energy in the process.
3. Doing so would drop its energy below the quantum level required for its existence.
4. Therefore, the electron would cease to exist.
5. But an electron cannot cease to exist without violating the laws of conservation of both charge and angular momentum.
6. Therefore, an electron cannot drop into the nucleus.

- The quantum of light is the photon. The energy of a photon is directly related to the frequency, f, of the photon. Specifically, $E = hf$. Here, "h" is the Planck constant (named for Max Planck). There is a minimal frequency below which no energy is observable and for which no photon can exist.

At this point, we can easily provide an intuitive derivation (not a proof) of the Heisenberg Uncertainty Principle. The uncertainty principle asserts that we can never determine a change of momentum $\triangle p$ and a change of position $\triangle x$ to within $\frac{\hbar}{2}$, where $\hbar$ is the reduced Planck constant.

Expressed as an equation, the Heisenberg uncertainty principle is:

$$\triangle x \, \triangle p \geq \frac{\hbar}{2},$$

Now, for our intuitive argument:

1. We'll begin with the well-known relationship between energy (E) and wave frequency (f).

$$E = hf \qquad \text{Equivalently,} \qquad h = E/f$$

But,

2. We know that when rest mass is negligible, $E = (pc)^2$, and we also know that for electromagnetic radiation, $f = c/\lambda$, where c is the speed of light, and lambda is its wavelength.

So,

$$h = \frac{pc}{\frac{c}{\lambda}} = p\lambda$$

3. We introduce here a naming convention that physicists find convenient, that of the "reduced" Planck constant, $\hbar$

$$\hbar = \frac{h}{2\pi}$$

Reduction converts the Planck constant into a measure in radians. We know that the angular frequency, ω is equal to $2\pi f$. So,

4.
$$E = hf = \hbar \times 2\pi f = \hbar\omega$$

Returning to our derived equation for the Planck constant,

5.
$$h = \frac{pc}{\frac{c}{\lambda}} = p\lambda$$

6. Dividing both sides by 2π we have,

$$\hbar = p \times \frac{\lambda}{2\pi}$$

$\frac{\lambda}{2\pi}$ is the reduced form of the wavelength, $\bar{\lambda}$ or the wavelength expressed as radians.

So, the minimal quanta of electromagnetic radiation requires,

7.
$$\hbar = p \times \frac{\lambda}{2\pi} = p\bar{\lambda}$$

This equation bears a striking resemblance to the Heisenberg uncertainty principle, which tells us that the closest precision we may attain for the measurement of a particle by position and momentum must be greater than a quantity determined by the Planck constant. Specifically,

$$\Delta p \times \Delta x \geq \hbar/2$$

Delta x is the minimal measurable distance for an observation and corresponds to the minimum wavelength of a photon of light (beneath which the photon would cease to exist). If we wanted to measure (i.e., observe) a wavelength, we would need to have a sampling probe that is smaller than the wavelength under observation. In the case of electromagnetic radiation, the sampling probe must be not more than half the wavelength of the object being measured.

Therefore, the two equations are related by a factor of two, and, for the minimal possible combination of momentum and wavelength (i.e., distance), $\hbar = p\bar{\lambda}$ is equivalent to a measurement uncertainty of $\Delta p \times \Delta x \geq \hbar/2$.[23]

[23] We will develop a mathematical proof of the uncertainty principle in the section titled "Proving the Heisenberg uncertainty principle".

1.8 Proving fundamental particles do not collide with one another

"The only difference between a problem and a solution is that people understand the solution".
—Charles Kettering

An electron is a point particle (i.e., a particle that has no size) that has mass (unlike the photon, which is a point particle without mass). What happens when two massive particles (such as electrons) approach one another on a collision course? Duh, they collide! . . . Or do they?

1. As the distance between the two particles diminishes, gravitational attraction increases.

   The force of gravity increases according to an inverse square formula. This means that the force of gravity on the particles increases four-fold as the distance between the masses is reduced by half. As the distance between the two point particles decreases to zero, the force of gravity increases to infinity.

   If this were so, then why doesn't the force of gravity increase to infinity every time two massive objects (like a tennis ball and a soccer ball) touch one another? When ordinary objects collide, their centers of gravity never touch. Their exteriors touch well before their centers of gravity come close enough to produce any significant gravitational effect. In the case of point particles having no size (i.e., no diameter), a collision would place centers of gravity of the two particles at one location, with no intervening space.

2. If the two electrons actually collided, an untenable situation would arise, wherein an infinite gravitational force would be exerted at one point in space.

   Such a scenario would likely result in a new Big Bang, replacing our universe with a brand new universe. The fact that our universe is more or less intact is strong evidence that fundamental particles do not collide. What does happen when two massive fundamental particles closely approach one another? During "near misses", short-lived low-energy particles (so-called virtual particles) enter existence and mediate a transfer of forces over very small distances. We won't be discussing this topic further, though it is of enduring interest to particle physicists.

3. Therefore, fundamental particles having mass cannot collide with one another.

We've just discussed why a massive fundamental particle, such as an electron, cannot collide with another massive fundamental particle. What about massless fundamental particles, such as photons? Can they collide? When we come to the discussion of the Pauli exclusion principle, we will find that the massless fundamental particles have the property of superpositioning. Two massless particles can exist at the same location at the same time, without colliding. They simply pass through one another in the manner of waves.

What about non-fundamental particles; can they collide with one another? Yes, protons and neutrons are composite particles composed of quarks and gluons. As such, they have dimensionality (size) and can approach one another without tearing apart the fabric of spacetime. Collisions between non-fundamental particles are commonplace.

1.9 Proving energy quantization results from boundary conditions on waves

A standing wave (also called stationary wave) is a wave that oscillates but does not move in the sense there is no succession of waves that pass a point. Examples of stationary waves are children playing with a jump rope, or a musician plucking a guitar string. Both ends of the rope and both ends of the guitar string are fixed, and the amplitude of the wave at either end is zero (because the ends are fixed and cannot raise their amplitudes). The wave oscillates up and down around an integer number of nodes having zero amplitude, with the two ends of the string being the first and the last of the zero amplitude nodes. Standing waves are always quantal insofar as there is an integer number of nodes holding an integer number of identical wavelengths.

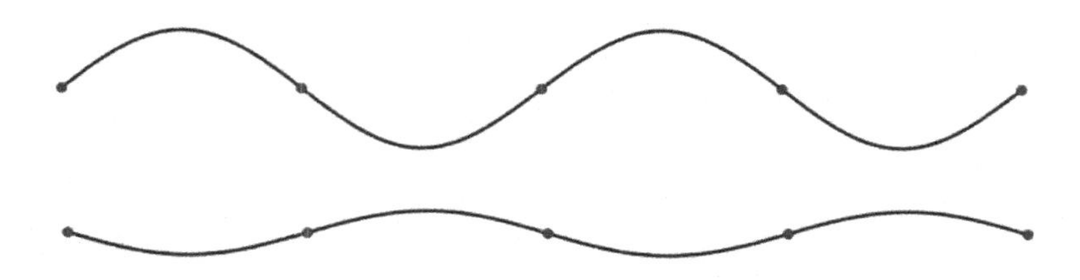

FIGURE 1.3: A vibrating string stretched between two fixed points will produce a standing wave. The locations of the nodes of zero amplitude stay fixed as the wave oscillates. Above, we see the wave at its peak. Below, we see the wave a moment later, when the amplitudes have fallen. Notice that the nodes of the wave do not move and that the fixed nodes at either end always have zero amplitude. There are five zero-amplitude nodes along the length of the oscillating string. The nodal configuration of a standing wave is a quantal system. The number of nodes is always an integer.

In our example of the stationary wave, we employ a system bounded by two points (the ends of the string), with the two points at either end of the string having zero amplitudes (i.e., the two ends of the string are fixed and bounded). We could have imagined a box, with waves bouncing inside. In this case, there would be imposed a zero wave amplitude at the borders of the box. Consequently, inside the box there would be an integer number of zero amplitude nodes. The higher the frequency of the waves inside the box, the greater the integer number of nodes.

The sound waves produced by a piano are standing waves having an integer number of evenly spaced nodes between the ends of the set of taut piano wires that constitute the piano. We get quantized notes because quantization is a feature of standing waves. When we produce sound using a vibrator that can be set to any frequency, the resulting sound waves cover a continuous range of frequencies and are not constrained to any quantal distribution (i.e., musical notes).

Suppose we choose to loop our string (e.g., jump rope or guitar string) into a circle, such that the two ends of the string are now touching one another, and the waves were confined to the circle. The same constraint applies. Waves traveling in a closed loop are standing waves, and the closed loop must have an integer number of nodes. If the situation were otherwise, and there were not an integer number of nodes, then with each rotation around the circle, the waves would move further out of sync, interfering with one another destructively, and destabilizing the vibrations in the string. We can imagine that orbits of electrons are standing waves akin to the plucked wires of a piano (but wrapped into a loop). All the electrons in an orbit are tuned to the same frequency; otherwise, they would interfere with one another as their repeated orbits put them further out of sync.

When an electron jumps from one orbit to another, it must have a frequency tuned to that of the other orbit. To do so, it must gain or lose energy that is itself quantized (i.e., constrained to a specific amount suitable for the orbital transition). The quantum of energy is exactly equal to the difference in the energies of the respective orbits. The energy that is gained or released is carried by photons. Therefore, the photons whose energy is absorbed or released in an orbital transition are characterized by electromagnetic radiation of a specific frequency. It is for this reason that the emission spectra of heated atomic elements always have the same set of thin emission bands, where each band represents quantal electromagnetic radiation released by one of many possible orbital transitions occurring in the heated atoms.

The requirement for the quantization of energy levels only applies to bound states, such as those that apply to orbiting electrons and piano wires. Unbound systems do not have quantized energies. Particles moving freely through space are not constrained to quantal frequencies. They might have any energy over a continuous scale from zero to near-infinity (infinity being the unattainable energy required to move a massive particle up to the speed of light).

1.10 Proving most of the observable energy of the universe is the cosmic microwave background

How did molecules first emerge from an early universe that contained only fundamental particles? It is thought that in the first moments after the Big Bang, fundamental particles formed, and these included photons, electrons, quarks, gluons, neutrinos, and their respective antiparticles. Eventually, quarks and gluons aggregated to form protons and neutrons.[24] Because protons and electrons have opposite charges, they attract, forming the first atoms (of Hydrogen and Helium), but for a very long time, the universe was far too hot for the earliest atoms to exist for any meaningful length of time. The prevailing temperature of the early universe provided kinetic energy that greatly exceeded the ionization energy of Hydrogen and Helium atoms (i.e., the prevailing high energy would instantly tear apart any newly formed atoms).[25] During this early hot period of the universe, photons moved the short distances from particle to particle, being repeatedly absorbed and emitted, but photons did not travel freely through space.

At about 380,000 years following the Big Bang, the universe cooled down to a comfortable 3,000 ° Kelvin, at which time the temperature of the universe could not ionize the electron in Hydrogen atoms, and every positively charged Hydrogen and Helium nucleus (these were the only two atomic nuclei synthesized at the time) quickly coupled with one or two negatively charged electrons (one electron for Hydrogen and two for Helium). These early atoms would be stable because the universe continued to cool as it expanded, and the temperature of empty space (i.e., the heat energy that vibrates atoms and molecules) never again reached a level sufficient to ionize atoms. For this reason, most of the atoms in our universe today are the original Hydrogen and Helium atoms that congealed nearly 13.5 billion years ago (380,000 years following the Big Bang).[26]

At the same moment that the first stable atoms formed, and the primordial hot plasma of bare particles ceased to exist, the photons found a clear path by which they could stream through the expanding universe. In a flash, the

---

[24]Protons and neutrons have nearly 2000 times the mass of electrons. Therefore most of what we encounter as matter (i.e., atoms and molecules) consist of protons and neutrons. But protons and neutrons consist of quarks (massive) and gluons (massless). Accordingly quarks and gluons account for more than 99% of the matter in our universe.

[25]The ionization energy is the energy required to pull electrons from atoms, leaving the behind an ionized atom whose positive charge (from protons) is not matched by negative charge (from electrons).

[26]Today, about 70% of the atoms of the universe are Hydrogen and 28% are Helium. Nucleosynthesis in stars accounts for only about 2% of the atoms in our universe. Most of the remaining 98% of matter (Hydrogen and Helium) are the original atoms dating back nearly to the Big Bang.

universe became transparent to light. The escaping light that filled the early universe remains with us today, and is known to us as the cosmic background radiation (or, more popularly, as the cosmic microwave background).[27]

When we aim our Earth-bound telescopes in any direction, we find that all of space is saturated with microwave radiation. Because this radiation has just about the same intensity wherever we look in the universe, we deduce that this is the cosmic background radiation that emanated through the cosmos 380,000 years following the Big Bang.[28] To this very day, the majority of the electromagnetic energy in our universe comes from cosmic background radiation. We cannot see the cosmic background radiation with our naked eyes because, as the universe has expanded, this primordial light has diminished in wavelength, dropping into the invisible microwave range. The cosmic background radiation currently peaks at a wavelength of about 1.9 mm and a frequency of 160.2 GigaHz. The distribution of the wavelength of the cosmic background radiation conforms to the black-body radiation curve, as we would expect with a non-reflective body heated to a temperature of 2.7° Kelvin.

Why do we conclude that most of the observable energy in our universe derives from the cosmic microwave background?[29]

1. Most of the universe is so-called empty space, containing nothing but sparsely distributed Hydrogen and Helium atoms. Stars and planets and condensed gases and the electromagnetic radiation they produce contribute insignificantly to the total energy in the observable universe.
2. The only observable source of energy, filling the entire universe, is the cosmic microwave background.
3. The measured temperature of outer space is 2.7 ° Kelvin.
4. The measured temperature of outer space is equal to the temperature provided by the cosmic microwave background (the only source of heat in an otherwise empty volume of space).
5. This would indicate that the source of the energy of the universe, capable of heating the universe to a temperature of 2.7 ° Kelvin, is the cosmic microwave background.

---

[27]Electromagnetic waves tell us little about the universe preceding the release of the cosmic background radiation (i.e., the first 380,000 years after the Big Bang). In contrast, the first gravity waves, in theory, were unaffected by absorption or scattering by primordial plasma. In principle, by studying gravity waves emitted at or near the time of the Big Bang, we can study the earliest moments of our universe, gaining insights unattainable with electromagnetic waves.

[28]The cosmic background radiation is remarkably uniform throughout space. Variations in its intensity are small and only the most sensitive instruments can detect local fluctuations.

[29]We exclude mention of dark energy, which is unobservable and hypothetical, but which is estimated to provide the majority of the energy of the universe.

Therefore,

6. The cosmic microwave background provides most of the observable energy of the universe.

Our claim that the cosmic microwave background is the major source of observable energy in the universe is surprising insofar as this universal background energy wasn't discovered until 1964 and is totally invisible to the naked eye.[30] As it so often happens, we must turn to our television sets for enlightenment. Readers who have not discarded their analog (cathode ray tube) televisions can watch the cosmic background radiation day or night, from any location on Earth. Turn on your television and detune the receiver (by turning to a channel that does not receive broadcast signals, or by disconnecting your antenna). Your picture tube will display "snow", a cacophony of flickering black and white dots completely filling the screen, produced in part by microwaves. It has been estimated that a few percent of the snow on analog TV sets comes from cosmic background radiation. We note that the cosmic microwave background was not always invisible. At some point in the history of our universe, the cosmic background radiation had a frequency that was well within the visible spectrum. Over time, the color of space changed from blue to red, and finally to black.

[30]The discovery of the cosmic microwave background is credited to Arno Penzias and Robert Wilson, for which they received the Nobel prize in 1978.

1.11 Proving chemical bonds do not exist

"Sometimes it appears to me that a bond between two atoms has become so real, so tangible, so familiar that I can almost see it. But then I awake with a little shock: for the chemical bond is not a real thing, it does not exist, and no one has ever seen or will ever see it. It is a figment of our imagination".
—Charles Coulson[31]

In science, figurative language often provides a familiar way of envisioning processes and relationships that would otherwise be difficult to fathom. We describe electrons orbiting around a nucleus, much like planets are seen to orbit around the sun. Easy to envision, but wrong. It would be more accurate to say that electrons occupy particular zones of probability in relation to the nucleus of the atom. The flawed concept of electrons in orbital shells, with sub-shells within the orbiting shells, helps us to understand the concept of valence that we apply to the outermost orbital shell of an atom. Once we've deluded ourselves into believing, at face value, the principles of valence, then the notion of physical-chemical bonds becomes credible. Every once in a while, we must remind ourselves that chemical bonds are not physical structures.

If chemical bonds existed, we would expect to see them. Instead, sensitive imaging techniques reveal the geometric configuration of atoms arranged in molecules, but the bonds themselves (i.e., the physical connections between the different atoms of the molecule) are nowhere to be seen. When molecules disintegrate, disassembling into individual atoms or short fragments of the original molecule, we do not see the tell-tale remnants of broken bonds. It is as though the molecular bonds never existed (as indeed they never did). A molecule is an association among nearby atoms that persists until such time as sufficient energy is made available to pull the constituent atoms apart. A stable molecule is one in which the energy to pull the constituent atoms apart is unavailable under current conditions. An unstable molecule is one for which the energy available to dissociate the atoms of the molecule is readily available. Given sufficient energy, all molecules will dissociate into individual atoms and all atoms will dissociate into a plasma consisting of nuclei and free electrons. At much higher energy levels, such as those seen at the first moments of the Big Bang, aggregate particles such as atomic nuclei and their constituent protons and neutrons, would dissociate into fundamental particles (i.e., quarks and gluons). What we call bonds are simply a shorthand description of an arrangement among the individual atoms that associate as a stable molecule.

[31] In Coulson CA. The contributions of wave mechanics to chemistry. Journal of the Chemical Society. 2:69-84, 1955.

To better understand the meaning of "stable" molecules, we first need to fully understand the meaning of "heat". The word "heat" only has meaning when applied to moving atoms. We cannot speak of heat in the context of empty space (i.e., space without matter) because when there are no vibrating atoms and molecules, there is no heat (there is only the energy of empty space). If we were to put a thermometer into empty space, it would register 2.7 ° Kelvin, but this is not the temperature of empty space. Rather, 2.7 ° Kelvin is the temperature of the thermometer in empty space. In fact, thermometers measure their own temperature, the temperature associated with the heat of the moving molecules in the thermometer. The energy of empty space causes the molecules of the thermometer to vibrate, producing heat, and the heat registers as the temperature of the thermometer.[32]

All molecules (i.e., geometric associations among atoms in close proximity) are stable at absolute zero temperature. The reason being that at absolute zero, there is no movement of the atoms of the molecule relative to one another. Everything is perfectly still at absolute zero. It is only when energy is introduced that the atoms of the molecule can begin to move, producing heat. When the heat is sufficiently high, the atoms in molecules will begin to dissociate.

The early universe was too hot for stable atoms to exist. Even when all the ingredients of an atom (protons, neutrons, and electrons) were available, high energy levels were sufficient to tear apart any nascent atoms. Over time, the universe cooled, and when the universe reached a temperature of about 3,000 ° Kelvin, the ambient energy was sufficiently low to permit the formation of stable Hydrogen and Helium atoms (see section titled "Proving that most of the observable energy of the universe comes from the cosmic microwave background" for additional background information on this topic). Atoms that bumped into other atoms could remain together if the available energy was insufficient to pull them apart. The associations between atoms that stood the best chance of persisting (as molecules) were achieved through interactions occurring between electrons in the outermost orbital shells of atoms.[33] Such interactions take the form of so-called ionic bonds or of covalent bond, but are more accurately simply geometric associations that are energetically stable until pulled apart. Today, spectroscopic studies of interstellar space indicate that in addition to Hydrogen and Helium, a variety

[32]At higher temperatures, when a thermometer is placed pressed against matter, such as a human tongue, the heat of the tongue equilibrates with the heat of the thermometer until the tongue molecules and thermometer molecules are vibrating at the same rate. We say that the thermometer is measuring the temperature of the tongue, but the thermometer is actually measuring its own temperature, equilibrated to the temperature of the tongue.

[33]Strictly speaking we should avoid terms such as outer shells and orbits, preferring to think in terms of electrons whose quantum states place them at probable locations that extend the furthest from the nucleus. For now, though, let's stay with familiar terminology.

of polycyclic aromatic hydrocarbons are present.[34] The same early complex molecules that formed about 380,000 years after the Big Bang, are still present throughout the universe. It should come as no surprise that polycyclic aromatic hydrocarbons play a major roles in the chemistry of life on Earth.

[34]see Hudgins DM, Bauschlicher CW, Allamandola LJ. Variations in the peak position of the 6.2 micron interstellar emission feature: a tracer of N in the interstellar polycyclic aromatic hydrocarbon population. Astrophysical Journal 632:316-332, 2005.

1.12 Proving chemical reactions are reversible

"A great truth is a truth whose opposite is also a truth".
—Thomas Mann

There is a very simple proof that all chemical reactions (in fact, all actions) are reversible, but the proof is contingent upon our acceptance that physicists have succeeded in developing all of the laws that describe physical actions. Here is the proof:

1. The universe operates under conservation laws that describe the fields, forces, particles, and all of their interactions.
2. The laws describe events that occur over increments of time.
3. The increments of time are directionally agnostic (i.e., our laws don't care whether time moves forward or backward).
4. Therefore physically predicted actions in forward time can just as easily occur in backward time.
5. Therefore, all physical actions are reversible in time.

Our conclusion, that all physical actions are reversible, seems to be incompatible with everyday observation. When we ignite a stick of wood, and the flames rise and the wood turns to ash, we can be fairly certain that there is no way to reverse the process. Aside from the seemingly irreconcilable differences between reactants and products, we must somehow contend with the second law of thermodynamics, which tells us that all reactions lead to an increase in entropy (never a decrease). In the case of the burning stick, heat is released from the fire, dissipating to space, beyond reconstruction, and irreversibly increasing the entropy of the universe.

Still, nature provides us with an abundance of sticks, and each stick must have been created through a series of reactions that acted upon the elemental constituents of wood (e.g., Carbon, Hydrogen, Oxygen, the principle elements present in ash). An important point in understanding the reversibility of any reaction is that time plays only a passive role. All reactions occur over some interval of time, but time does not make the reaction occur. What makes the reaction occur are the physical laws acting upon substrates and products. These physical laws are time-agnostic and are valid whether time moves in the forward direction or the backward direction.

To the detriment of our "proof" of the reversibility of physical actions, we must contend with the collapse of the Schrödinger wave equation, a phenomenon that we will describe in some depth later in this book (see section titled "Proving that the collapse of the wave function invalidates measurement"). In brief, particles behave much like waves, and beams of fundamental particles emerging through adjacent slits will produce an interference pattern such as we would expect to see with any self-respecting set of waves. However, when a sensor is placed in the path of the light emerging from the slit, the interference pattern (a wave phenomenon) abruptly stops, yielding a pattern that would be expected from a beam of ejected particles that have lost their wave-like property. This is the fabled collapse of the Schrödinger wave equation, and the wave equation supposedly describes the evolution of all quantum objects. The wave equation is, like all of our laws of physics, time-agnostic. The wave equation describes a universe in which particles occupy all possible states, at once. When the wave equation collapses, one particular state is "chosen", accounting for the apparent shift from wave-to-particle. This choice of state is, as far as we can tell, irreversible. We cannot move backward from one specific state back to its predecessor state because the predecessor state represented all possible states, at once. Hence, the collapse of the wave equation is time-irreversible and would seem to invalidate our "proof" that all physical actions are reversible. Let's not despair. Nobody can claim to truly understand the collapse of the wave function. What the collapse of the wave function tells us about the reversibility of chemical reactions remains an unsolved puzzle.

1.13 Proving there is an upper limit to Periodic Table

There are just 94 naturally occurring elements. Heavier elements have been formed in laboratories, but the man-made elements are unstable and pass out of existence after mere fractions of a second. Is there a theoretically determined upper limit to the number of elements?

It turns out that as the mass of an element increases, the velocity of its electrons must also increase. The reason being that as the mass of the nucleus increases, the innermost electrons are drawn into a smaller orbit, and the velocity of the electrons must increase to conserve angular momentum. The speed of innermost orbital electrons for any element is determined by the element's atomic number. The equation describing the relation between electron speed and an element's atomic number is:

$$v_{relativistic} = Z/137$$

V is the velocity of the inner orbital electron relative to the speed of light, Z is the atomic number of the element, and 137 is the dimensionless fundamental constant known as the inverse of the fine structure constant.[35] A ground state electron in gold (Z = 79) will travel at 58% of the speed of light. We can see that when the atomic number of the element is 137, the electrons in the inner orbit would travel at the speed of light, a physical impossibility. **Therefore, elements with atomic number 137 or higher cannot exist.** [36] Hence, there can never be an element of atomic number 137 or higher. [37]

Regardless of which theoretical limit is chosen for the highest possible atomic number of an element, it is clear that some limit must exist insofar as the mass of the electrons will approach infinity as their speed approaches the speed of light.

---

[35]Early in the 20th century, the fine structure of the Hydrogen atom's emission spectrum was observed to produce two narrowly separated lines, where only one line was predicted by the Bohr theory of the atom. The fine structural separation of spectral lines was quantified by Arnold Sommerfeld as a dimensionless value, 1/137, and has been called the fine structure constant or Sommerfeld's constant (1916). Since then, numerous interpretations of the physical meaning of the fine structure constant have been offered. Values for the constants have been more precisely established from experimental measurements and from predictions based on quantum electrodynamics theory.

[36]The equation for the relativistic mass of particles (including orbiting electrons) moving at relativistic speeds is:

$$m_{relativistic} = m \Big/ \sqrt{1 - \frac{v^2_{relativistic}}{c^2}}$$

As $v \to c$, the mass of the object approaches infinity.

[37]Various scientists, using different models, have arrived at other upper limits for the Periodic Table. A nontechnical discussion of the topic is found in Philip Ball's essay, "Would element 137 really spell the end of the Periodic Table?", in Chemistry World. November, 2010.

1.14 Proving electron capture (by protons) does occur

Electron capture refers to the process by which an electron interacts with a proton to produce a neutron plus a neutrino.[38, 39]

proton + electron → neutron + neutrino

How do we know that electron capture occurs? The existence of neutron stars composed almost entirely of bare neutrons, is fairly strong evidence for electron capture. Here is how neutron stars are formed:

1. When the fusion process of a star comes to an end, the outward force from the energy released by fusion can no longer counteract the attractive force of gravity.
2. The proton-rich atoms in a star, particularly Hydrogen and Helium, are compressed by gravity, reducing the spaces between atoms and pushing electrons into any available lower orbits.
3. As electrons are compacted into low orbits, the Pauli exclusion principle exerts a so-called degeneracy pressure prohibiting any two electrons from occupying the same quantum state.
4. The pressure mounts to the point where sufficient energy is available for electron capture. Accordingly, protons and electrons are replaced by neutrons and neutrinos.
5. Eventually, the compressed star is replaced almost entirely by neutrons. As neutrons are created, neutrinos are released (one neutron for each captured electron).
6. The result is a neutron star, having a diameter of just a few miles, and exhibiting all of the physical characteristics that cosmologists would expect to find in a hyperdense and stable collection of neutrons.

[38]As a point of clarification, it is important to understand that the electron does not merge into the proton to yield a neutron that expels a neutrino. In all fundamental particle interactions (including fundamental particles such as electrons, neutrinos, and quarks), interactions yield the creation of new particles. In this case, the neutrino that is expelled in the electron capture reaction did not exist until the reaction occurred; the electron ceased to exist when the reaction occurred.

[39]A proton is replaced by a neutron when an "up" quark ceases to exist and a "down" quark comes into existence.

7. The evolution of neutron stars simply would not be possible if protons and electrons were not replaced by neutrons, thus demonstrating the existence of electron capture.[40]

What happens to all of those neutrinos produced in the electron capture reaction? Neutrinos are massless, uncharged particles that travel at the speed of light; they escape into space. In so doing, they carry energy away from the neutron star, thus causing the neutron star to rapidly cool, a phenomenon that may occur over a few decades, and which is observable by astronomers here on Earth. The cooling process of evolving neutron stars is one additional observation that supports the electron capture phenomenon.

Because quantum events are time-reversible, we might expect to find a reverse electron capture reaction. As it happens, the reversal of electron capture is neutron decay, an easily observable phenomenon represented by the following:

neutron $\rightarrow$ proton + electron + antineutrino

Insofar as electron capture requires energy, we can expect that the neutron decay reaction (i.e., electron capture in reverse) will release energy. Neutron decay occurs here on Earth, serving as one of the principle phenomena tracked by a commonly employed technique to measure the age (i.e., the time of formation) of rocks containing crystals, such as zircon. Zircon is a silicate crystal that is distributed widely through the crust of Earth. When we look closely at rocks containing zircon, we find that atoms of lead can be found, locked solidly within the silicate crystal, replacing Uranium atoms which are the preferred constituents of the zircon crystal lattice. The presence of lead in zircon had been something of a mystery, as we know that zircon, as it forms, excludes lead molecules from its matrix. The mystery is analogous to one of those whodunits wherein the murder victim was alone, in a room locked from the inside. How did the murderer commit the crime? How did lead atoms insert themselves into the crystal lattice of zircon?

In the case of zircon, Lead atoms gained access to the zircon crystal matrix via transmutation of Uranium. Radioactive isotopes of Uranium, entombed in zircon, slowly decay, eventually yielding lead. In the process of decaying, neutrons are replaced by protons, changing leads to elements having fewer neutrons and additional protons. The process releases energy, in focally damaging the crystal and producing visible micro-fractures, a process known as metamictization. As time passes, the number of lead atoms in zircon increases, along with the number of microfractures. Using a low-power

[40]As one might expect, an atom having no electrons cannot participate in electron capture. Most atoms have several different decay pathways, electron capture being just one of them. In the case of Rubidium-83, electron capture is the only decay pathway available. If Rubidium-83 atoms were stripped of their electrons, there would be no electron capture and no radioactive decay. In such a case, Rubidium-83 atoms would be stable for eternity.

microscope or even a humble hand lens, an astute mineralogist can detect the presence of radioactive decay in rocks, and the traces of released energy.

If we know the concentration of the various Uranium isotopes that were incorporated into zircon, as it formed, and if we know the rate of radioactive decay of these isotopes, and if we know the concentration of lead molecules in the present-day zircon, then we can calculate the date that the zircon formed. Doing so has yielded a great deal of information about the process of crustal rock formation. Of the rocks that have been sampled for radioactive dating, the oldest seem to be somewhat greater than 4 billion years old; an indication that the Earth was sufficiently cool to form solid rocks within about a half billion years of its formation. Moreover, these findings confirm that some present-day rocks are incredibly ancient, enduring for 4 billion years, or more.

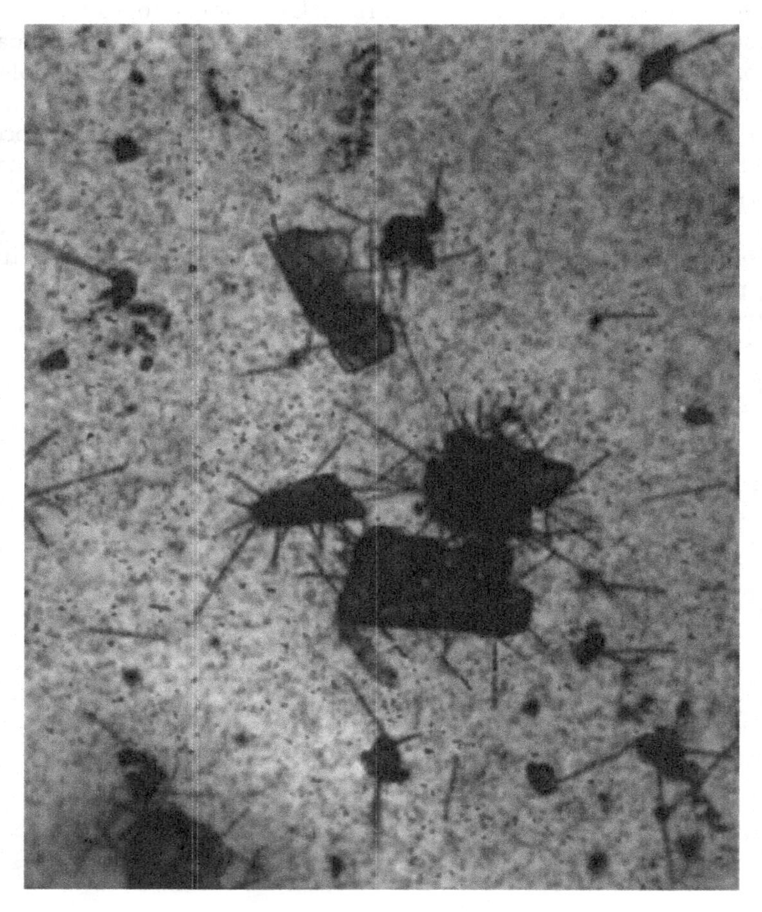

FIGURE 1.4: Photograph of alpha tracks in rock, emitted by Uranium molecules and traveling through a silver-based photoemulsion, developed after a 5-day exposure. The black (i.e., developed silver grains) areas with radiating lines are sites of Uranium molecules or Uranium daughter products. Each radiating line is an alpha particle track. If we were to prepare another photo-emulsion image of the same rock, after we wait a few hundred million years, the number of alpha tracks on the image would be fewer, as a portion of the Uranium atoms would have decayed into non-radioactive lead. Source: Public domain image, U.S. Department of the Interior, 1951.

1.15 Proving fundamental particles do not decay

Decay is all around us, affecting big and small objects on Earth and in space: Stars have a long but limited lifetime, orbits decay, giant sequoia trees die and decay, animals die and decay, radioactive isotopes decay, and neutrons decay. We might expect that fundamental particles must, like all other things, succumb to the after-effects of a self-limiting lifespan. Actually, we have good reason to believe that fundamental particles such as electrons, neutrinos and quarks, are endowed with eternal stability and do not decay.[41]

We posit the following:

- A particle may transition to a lower energy state.
- A particle that has no lower energy state to which it can transition is stable and cannot decay.
- Such particles, in the absence of intervening events, are eternal.

It may surprise you, but most of the electrons, protons, and photons that were created soon after the Big Bang are still in existence and are likely to remain in existence for a very long time. For example, protons are estimated to have a lifespan of at least 10^{34} years. Electrons are thought to have a mean lifetime exceeding 10^{24} years. At this point, readers may interject that every particle has a well-described path to destruction: electrons can be annihilated by positrons; protons can be annihilated by anti-protons; photons can be absorbed by opaque matter; protons and electrons can collapse together to form neutrons (as we see in neutron star formation). It's all true, but none of these aforementioned processes are true examples of decay; scientists apply the term "decay" carelessly to just about any reaction in which a particle ceases to exist. In addition to **not** being decay processes, they are also rare. Nearly all of the matter in our current universe is found in the persistent remains of interstellar Hydrogen and helium formed a few hundred thousand years after the Big Bang. The cosmic background radiation accounts for most of the light traveling freely through the universe and was released from a primordial hot plasma at the same moment when atomic Hydrogen and Helium atoms were created.

Some of the fundamental particles have very simple explanations for their stability.

- Electrons cannot decay because they carry charge, which is conserved.

[41] The word "decay", as used in this section, only applies to processes observed in particles that are not the result of forces external to the particle. For the purposes of our discussion, results of particle interactions in which one particle ceases to exist and is replaced by other particles, are not "decay" reactions, though they are sometimes referred to as such in the literature.

An electron cannot simply "decay" without causing a drop in the total charge present in the universe, and this would be a violation of the conservation of energy.[42]

- Protons, like electrons, carry charge, and cannot decay without violating a conservation law.

  Neutrons, unlike protons, do not carry charge and can decay. A free neutron decays quickly, with an expected lifetime of under 15 minutes.[43] Because a neutron carries no charge, neutron decay does not violate charge conservation.[44]

- Photons, being timeless, cannot decay inasmuch as decay is always a **time-dependent** phenomenon.

  Photons are the quanta of electromagnetic waves. Photons live in the "eternal now" attained by any particle traveling at the speed of light. For a photon, one moment in time is not observably different from any other moment in time. Accordingly, at light-speed, internal events that play out over time (such as decay) do not occur (they just can't find the time).

[42]An exception would be the balanced annihilation of one electron and one oppositely charged positron. As mentioned, an electron-positron annihilation is not an example of decay in that it is not spontaneous and involves more than one particle.

[43]For a discussion of neutron lifespan, see Pieter Mumm's article, "Resolving the neutron lifetime puzzle", in Science 360:605-606, 2018.

[44]When a neutron decays, it yields a proton (positive charge) plus an electron (negative charge) plus an anti-neutrino. Some forms of neutron decay may yield high-energy gamma radiation (i.e., high-frequency electromagnetic waves) and highly energetic electrons. Because a great deal of energy may be released when neutrons decay, the reverse reaction (neutron formation) is energetically unfavorable.

1.16 Proving the correspondence principle does not always apply

"Our similarities are different".
—Yogi Berra

The classical limit (also known as correspondence limit or correspondence principle) holds that the results predicted by quantum physics must be the same (to within a close approximation) as the results predicted with classical physics, when tested on a macroscopic scale.

The proof of the correspondence principle is based on a simple-minded extrapolation:

1. All of the fundamental particles of the universe are the quantal units of physical forces.
2. Macroscopic matter and electromagnetic waves are composed of quantal units and are both subject to the same physical laws that govern the quantum world.[45]
3. Fundamental particles and macroscopic matter have the same degrees of freedom in space and undergo the same types of movements in space (e.g., translation, rotation) and time (translation), and therefore have the same limited set of possible event outcomes (i.e., they can change position, their time, and their momentum and not much else).
4. Though it is impossible to calculate the interactions of every fundamental particle in macroscopic matter (there are just too many particles to contemplate), it is possible to use the laws of averaging to predict the behavior of matter, based upon our knowledge of the interactions among individual fundamental particles.

   Therefore,

5. Predictions based upon our knowledge of the quantum world should apply to the macroscopic world, to the accuracy that we might find when we apply statistical laws of averaging to large numbers of quantal objects.

[45] As a caveat, we should comment that gravity does not seem to fit into a quantal model. Gravity, as we understand it today, is a feature of the structure of space, and, to the best of our understanding, gravity affects everything within space in the same manner (i.e., individual particles, asteroids, and beams of light all conform to the general laws of gravity. Thus, for the special case of gravity, there is no need to have a correspondence principle insofar as the quantal world and the macroscopic world are functionally equivalent.

At first blush, the proof seems straightforward, but after a moment's thought, we see that some of the symmetry of the quantal world is lost on the macroscopic world, and we cannot say how these differences might alter our predictions. For example, there seems to be time symmetry for interactions among individual quantum particles. Any interactions occurring in one direction of time could just as well occur in the opposite direction. This is not the case for macroscopic interactions. We can send a truck over a cliff and watch it smash on the rocks below, scattering a thousand shreds of metal onto the landscape. We cannot expect those shreds of metal to reassemble and launch back onto the cliff edge. Density is another feature of the macroscopic world that does not seem to apply to the quantum world. We can build a sturdy two-foot model bridge from toothpicks, but a bridge that spans a two-mile gorge should not rely on toothpick-based technology.

In addition, there are properties of the macroscopic world that are the direct product of individual quantal or relativistic events happening within matter and are not predictable based upon averaging the behavior of large populations of quantal objects. Lasers, atom bombs, field effect transistors, quantum tunneling microscopes, fluorescence, and many other macroscopic phenomenon are a few examples where rare quantal events (not averaged events of countless particles) produce quantal outcomes that intrude themselves into the macroscopic world. Within individual atoms, processes occur that are relativistic, having no correspondence with processes observed when we move visible objects through space and time.

For example,

- The ground state electrons of all heavy elements move at relativistic speeds. For example, the electrons in gold atoms travel at 58% of the speed of light. Atoms, aside from being very small, are intrinsically relativistic and can never achieve a non-relativistic mode.
- The mass of individual protons and neutrons is about 2,000 times the mass of electrons. Nearly all of the mass of protons and neutrons is accounted for by the confinement energy associated with the gluon-quark interaction (i.e., the energy required to confine the strong force of quark-quark interactions). The rest-mass of protons and neutrons is negligible. Hence, the mass of protons and neutrons cannot be compared directly to masses measured in the classical world (although classical mass is composed almost entirely of protons and neutrons).
- Classical physics does not tell us why the element Gold is golden, or why the element Silver has the color of silver, or why Lead is as dense as it is, or why ions flow in Lead-acid batteries; but all these macroscopic properties are best understood as the result of fast-moving orbital

electrons conforming to Pauli's exclusion principle. [46] There is no classical explanation for the behavior of these classical objects.

In these examples, classical physics offers no explanation. Only quantum physics guides our understanding of macroscopic physical reality.

In summary, the correspondence principle asserts that the macroscopic world is the quantum world, writ large. The quantum laws pertain to single particles. It is tempting to believe that an emergent system "averages out" the quantum effects to yield the laws of classical physics. The correspondence principle holds true for much of physics. Nonetheless, the quantum world is known to yield macroscopic effects that classical physics does not explain or predict, and for which there is no correspondence.

[46]For further information, read: Pyykko P, Desclaux JP. Relativity and the periodic system of elements. Acc Chem res 12:276-281, 1979; Pitzer, KS. Relativistic effects on chemical properties. Accounts of Chemical Research 12:271-276, 1979; and Ahuja R, Blomqvist A, Larsson P, Pyykko P, Zaleski-Ejgierd P. Relativity and the lead-acid battery. Physical Review Letters. 106:018301, 2011.

2

Waves and Forces

> *"Light is a derived concept: it arises as a ripple of the electromagnetic field".*
> —David Tong

This chapter describes how fundamental particles can be conceived as wave packets. In the course of this chapter, proof is offered that waves, their superpositions, and their rotations can all be easily expressed in terms of Euler's formula (which is itself proven in the text). A short derivation of the Schrödinger wave equation is also provided. Other topics in this chapter include: a proof that fundamental particles and wave packets are equivalent, a proof that the speed of light in a vacuum is constant, a proof of Einstein's famous equation that energy equals mass times the speed of light squared, and proof that photons are massless but that massless light carries energy and momentum. An understanding of physical waves leads to an understanding of how all known physical interactions unfold.

DOI: 10.1201/9781003516378-2

2.1 Proving all objects fall at the same rate, regardless of their mass

"Almost all of gravity is just that clocks run at different speeds in different places."

—Jonathan Oppenheim, as quoted in the Quanta Magazine YouTube offering titled, "A Bet Against Quantum Gravity".

Galileo Galilei famously claimed that all objects fall at the same speed, regardless of their respective weights. Every student learns that Galileo proved that objects of different weights fall at the same speed during a public display conducted at the tip of the leaning tower of Pisa. Fewer people know that Galileo's famous experiment in Pisa merely confirmed what Galileo had already deduced, from a simple gedanken experiment.[1]

1. Imagine that you have a 2-pound ball and an 8-pound ball, connected by a string.
2. Let's pretend that heavy objects fall faster than light objects.
3. In this case, the 8-pound ball would drop faster than the 2-pound ball.

   But,

4. If this were so, then when the connected balls were dropped, the 2-pound ball, traveling slower than the 8-pound ball, would put a drag on the descent on the 8-pound ball; slowing its fall.

   Hence,

5. A 10-pound composite weight would travel slower than a single, 8-pound weight
6. This tells us that the original premise, that heavier objects fall faster than lighter objects, must be false.

Had we begun with the opposite assumption, that lighter objects fall faster than heavier objects, then the fast-traveling 2-pound weight would speed the descent of the attached 8-pound weight. Hence, the composite 10-pound weight would travel faster than the component 8-pound weight. Again, the

[1]Gedanken is the German word for "thought." A gedanken experiment is one in which the scientist imagines a situation and its outcome, without resorting to any physical construction of a scientific trial. Albert Einstein, a consummate theoretician, was fond of inventing imaginary scenarios, and his use of the term "gedanken experiments" has done much to popularize the concept. The scientific literature contains many descriptions of gedanken experiments that have led to fundamental breakthroughs in our understanding of the natural world.

original premise would be contradicted. Only if both connected weights, 2-pound and 8-pound, fell at the same rate, could this contradiction be avoided.

Galileo understood that the experiment at the Tower of Pisa was basically all for show. Its outcome was pre-determined by theory. Unfortunately for Galileo, real-world experiments do not always work as intended. When the balls were dropped from the Tower, the larger ball beat the smaller ball to the ground, by a smidgen. We now know that the difference is due to air resistance. When an equivalent experiment is conducted in a vacuum, the advantage of the heavier ball is lost, and the two balls of different weights drop at the same speed. At the time, Galileo's skeptical colleagues showed no tolerance for imprecision. The heavier ball beat the smaller ball to the ground; that fact was sufficient to conclude that heavy objects fall faster than light objects. His detractors were wrong, of course, but Galileo learned an important lesson in the politics of science; that experimental results are easily misinterpreted.

Had Galileo been just a bit more insightful, he may have taken his conclusion one step further, preempting one of Einstein's most important theories. Galileo had correctly reasoned that all objects, regardless of the type of atoms that compose the object, fall at the same rate under the influence of gravity. From this assertion, Galileo could have inferred that gravity is not a physical force. How so? Forces that act on objects must contend with the physical properties of the object, such as mass, molecular composition, and the properties of the individual molecules that compose the object. Much of physics can be reduced to the mathematical description of the interplay between forces and atoms. As we think about gravity, the physical properties of objects seem irrelevant. When we drop two objects, they fall at the same rate, regardless of their respective masses, or of the chemical composition of the objects. Moreover, gravity seems to be everywhere, at once. Whether you drop two objects from the Leaning Tower of Pisa, or the mountains of mars, the objects will fall as one, although the speed of the fall will be somewhat different on Earth than on mars. We cannot shield an object from the effects of gravity. It is as though gravity is a fundamental property of the universe. Hence, Galileo, back in the year 1590, could have inferred that gravity is not a force at all; that it must be part of the fabric of space and time. About 315 years later, Albert Einstein concluded that gravity is not a force; it is a curvature in space-time. When objects fall, they are not reacting to a force; they are slipping along a curve in space and time.[2]

[2]Galileo pondered some of the same problems that attracted the attention of Albert Einstein. Galileo had developed how own theory of relativity, to account for the differences of measurements conducted in different frames of reference. The piece of the puzzle known to Einstein but unknown to Galileo, was that the measured speed of light is the same in all frames of reference. Had Galileo been apprised of this fact, he may have deduced the laws of special relativity 300 years earlier than Einstein.

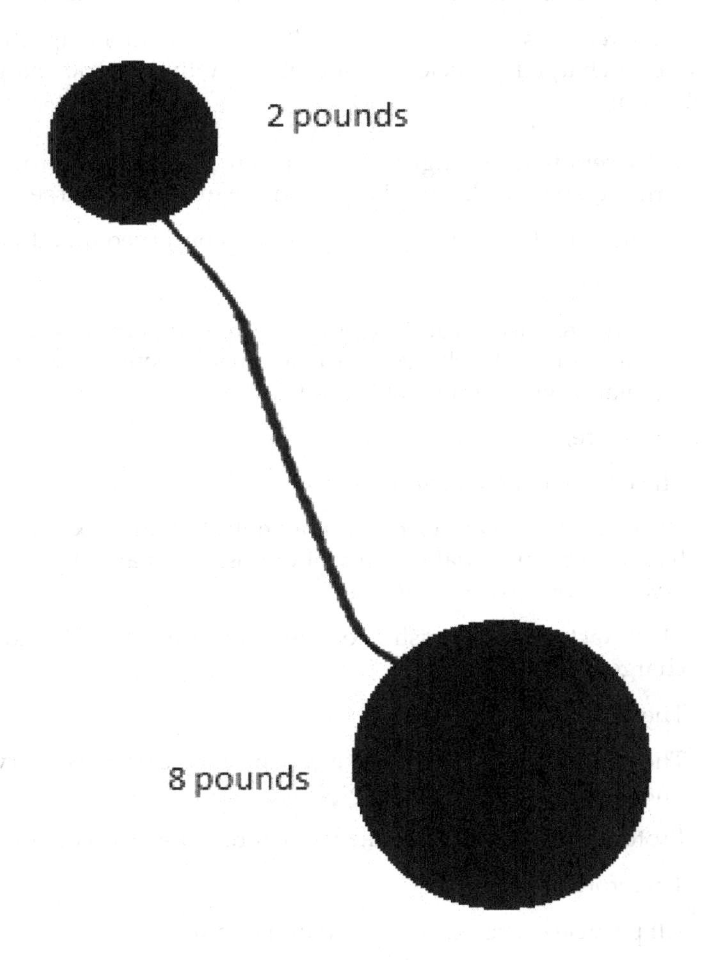

FIGURE 2.1: Galileo deduced, before his experiment at the Tower of Pisa, that all objects fall at the same speed, regardless of weight. The premise of his argument is that any object is composed of objects having a lesser weight. In the case of the illustration, a 10-pound weight is equivalent to an 8 pound weight and a 2-pound weight, joined together.

2.2 Proving all electrons carry the identical charge

There are various approaches to the issue of the quantization of charge.[3] We'll present two different arguments that lead to the same conclusion:

Argument 1 requires us to accept a well-known relationship: that the momentum of a charged particle is determined by the electric charge and the magnetic charge.

1. We accept that the angular momentum of a charged particle is determined by the electric charge and the magnetic charge.
2. Momentum (angular or otherwise) is a conserved quantity.

   Therefore,
3. The charge (electric and magnetic), which determines the angular momentum of the charged particle must be conserved (so that the angular momentum is conserved).

   Therefore,
4. The charge of an electron is conserved.
5. We know that every electron in a neutral atom is exactly counterbalanced by an equal number of oppositely charged protons (otherwise, atoms would not exist).
6. The exact 1:1 relationship between electron charge and proton charge is quantal.

   Therefore,
7. The charge of a proton is, like that of the electron, conserved and quantized.
8. Protons and electrons are the two charged particles of atoms.[4]

   Therefore,
9. All particle charge is conserved and quantized.

---

[3]Physicists understand charge conservation and quantization in terms of gauge symmetry, yielding a continuity equation that applies to charge, from which the Maxwell's equations arise. We won't delve.

[4]To be fair, the $+1$ charge of the proton comes from a complement of fundamental particles known as quarks. Each quark within a proton has a specific fraction of charge, the candidate compositions being two "up" quarks and one "down" quark (or perhaps four quarks and one anti-quark). The "up" quarks each carry a charge of $+2/3$ and the down quark carries a charge of $-1/3$, yielding a proton with a charge of $+1$.

Argument 2 is a bit more sophisticated than Argument 1 insofar as it introduces Noether's theorem. We will discuss Noether's theorem in detail in the section titled "Proving Noether's theorem", where we establish the relationship between observed symmetries and conservation laws.

1. There is an invariance associated with electric fields.

   The invariance of electric fields is the fixed relationship between electric and magnetic potentials (that account for the motion of electromagnetic waves).

2. Whenever there is a field invariance, there is a symmetry and a conserved element (Noether's theorem).

   At this point, all we need to know is that in the case of the electromagnetic field, the conserved element is charge.

3. Fundamental particles of a kind are identical to one another (see section titled "Proving the identicality of fundamental particles of a kind").

4. Fundamental particles are defined by their properties and the identicality of the fundamental particles (of a kind) requires that their properties must be identical.

   Therefore,

5. If the conserved charge is distributed among fundamental particles of a kind, then the charge must be the same among all identical particles of the same kind.

The two charge-carrying particles are the electron (which we call negatively charged), and the proton (positively charged). Because charge is conserved and because the identicality of charge-carrying particles constrain the amount of charge to the same quantity, we infer that every electron carries the exact same charge as every other electron. Protons, being composed of quarks and gluons, are not fundamental particles, but all protons are identical to one another, and every proton carries the exact same charge as every other proton.

Given that every electron has a charge of -1, and every proton has a charge of $+1$, it would seem that charge is quantized in the sense that it always has an integer value. As an aside, the Standard Model of quantum physics does not require charge to be quantized. We infer the quantal nature of charge by knowing that charge is conserved (i.e., positive and negative charges are balanced), and that charged fundamental particles of a kind all have identical charge.

2.3 Proving light is quantized

A quick glance at the electromagnetic spectrum is sufficient to tell us that the frequency range of electromagnetic waves is continuous and broad (i.e., non-quantized). We also know that the energy carried by an electromagnetic wave is determined by the frequency of the wave times a constant (i.e., $E = h\nu$). Therefore, the energy produced by light is also continuous and unquantized. Therefore, it would seem that our observations indicate that light itself is not quantized.

Well, perhaps we have overlooked the fact that quantized systems may yield continuous effects. Sometimes, we need to tease out the quantum from the background noise.

Here is a very simple argument that establishes the quantum nature of light:

1. Light sensors can be constructed that detect individual photons of light.

   That is to say that the electromagnetic wave loses its wave properties when it interacts with the sensor, and the sensor collects individual particles of light (i.e., photons). [5].

2. But we have proven that fundamental particles of a kind are absolutely identical to one another.

3. The fundamental particle of light is the photon, and all photons are absolutely identical to one another.

4. Identicality is a form of quantization.

   We can never find one-half or one-third of a photon, and a ray of light must contain an interger number of photons.

5. Therefore, electromagnetic waves, being composed of quanta, are quantized.

Albert Einstein produced a thorough and elegant proof of the quantization of light based upon observations of the photoelectric effect [6] This work has been covered in hundreds of textbooks and does not bear scrutiny here. Suffice it to summarize the points relevant to our discussion:

[5]The so-called collapse of the wave function occurs when light interacts with its environment (e.g., matter), and has been the subject of intense speculation extending over the past century

[6]Einstein's work, published in 1905, earned him the Nobel prize, in 1921.

- The quantum of light is the photon
- The intensity of light is determined by the density of photons

  For a given frequency of light, the only way to increase the intensity of the light is to recruit more photons (i.e., more quanta oscillating at the given frequency). This would be equivalent to having lots of flashlights all operating at the same frequency with which you can increase the total intensity by combining the beams of some of the other flashlights.

- The energy of light is determined by the frequency of oscillation of the photons

  Here is where the non-quantized factor enters the fray. The frequency range of light is continuous so that the energy (equal to a constant times the frequency) is also continuous. The photons, which are quantized, can be excited to any (non-quantized) frequency.

2.4 Proving photon creation must occur in pairs

"There are only 10 types of people;
those who think in binary, and those who do not".
—Often repeated, of unknown origin

1. Photons have spin.

   Therefore,

2. The creation of a singleton photon would violate the law of conservation of angular momentum (by creating an unbalanced spin).

   Therefore,

3. We would expect that every newly created photon must be accompanied by an identical photon having an opposite spin (conserving the angular momentum of the universe).

4. By the same token, photon annihilation must occur in pairs.

Experimentally, we find this to be the case (i.e., photons are created as pairs). The same argument applies to other types of fundamental particles.

2.5 Proving the inverse square law

The inverse square law states that the force emanating from a point source diminishes as the inverse square of the distance from the source. This law arises as a consequence of two conditions:

1. That force moving through empty space is conserved. If you have a force (e.g., light, or gravity) that emanates from a point (the so-called point source) then the force generated will be conserved throughout space.
2. The spread of force from a point source is spacetime invariant, so that the force is distributed (diluted) equally through all points extending from the source.

Let's imagine a point source. The force emanating through empty space from the point source can be envisioned as an enlarging sphere, the sphere representing the total symmetry of the expanding force arriving with equal strength at all points. Due to the conservation of force, the total force moving through a spherical shell at radius "r" will the equal to the total force moving through a spherical shell at radius "2r" and at radius "4r" and so on.

The area of a 3-dimensional sphere is $4\pi r^2$. This means that the total force moving through a sphere of radius r will be equally distributed over the area $4\pi r^2$. As the force expands outward, the same total force moving through a sphere of radius 2r will be distributed over an area of $4\pi(2r)^2$, or 4 times the total area of the smaller radius. This means that the passing force measured per unit surface area is reduced by 1/4, when we double the distance from the original point source (i.e., as an inverse square relationship).

Generalizing, the expanding force reduces as the inverse square of the distance from the emanating point source, in 3-dimensional space. We could extend the same line of reasoning to any n-dimensional space, as shown:

$$I \propto \frac{1}{r^{n-1}}$$

2.6 Proving waves can be expressed in terms of Euler's formula

A standing wave (also called stationary wave) is a wave that oscillates but does not move in the sense there is no succession of waves that pass a point. Examples of stationary waves are children playing with a jump rope, or a musician plucking a guitar string. Both ends of the rope and both ends of the guitar string are fixed, and the amplitude of the wave at either end is zero (because the ends are fixed and cannot raise their amplitudes). The wave oscillates up and down around an integer number of nodes having zero amplitude, with the two ends of the string being the first and the last of the zero amplitude nodes. Standing waves are always quantal insofar as there is an integer number of nodes holding an integer number of waves, with each wave oscillation being identical to every other wave oscillation.

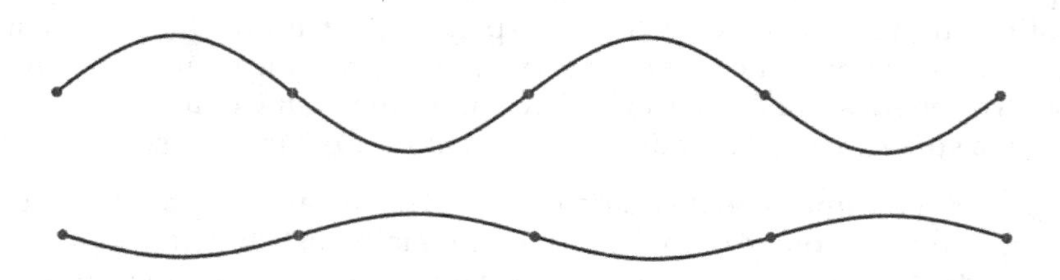

FIGURE 2.2: A vibrating string stretched between two fixed points will produce a standing wave. The locations of the nodes of zero amplitude stay fixed as the wave oscillates. Above, we see the wave at its peak. Below, we see the wave a moment later, when the amplitudes have fallen. Notice that the nodes of the wave do not move and that the fixed nodes at either end always have zero amplitude. There are five zero-amplitude nodes along the length of the oscillating string. The nodal configuration of a standing wave is a quantal system, with the number of waves described by an integer.

Standing wave oscillation can be described succinctly as e^{ikx}, where k is the wave number, the number of wavelengths in one revolution (i.e., $k = 2\pi/\lambda$, where λ is wavelength), i is the unit imaginary number, and x is a distance along the wave). It is far from obvious that e^{ikx} has much to do with waves, until we apply Euler's formula (see section titled "Proving Euler's formula and Euler's identity").[7, 8]

$$e^{ikx} = cos(kx) + isin(kx)$$

We see that the function can be represented as the sum of cosines and sines, both of which oscillate like waves, giving us a wave function. The wave function can be compactly expressed as Euler's formula. We're done!

While we're at it, we can apply the DeMoivre identity (see section titled "Proving the DeMoivre identity") to find, for the right side of our equation:

$$cos(kx) + isin(kx) = (cos(x) + isin(x))^k$$

The wave is provided with a real component, $cos(kx)$, and an imaginary component, $isin(x)$. At present, we focus on the real component. Substituting $2\pi/\lambda$ for k we have

$$cos(kx) = cos(\frac{2\pi}{\lambda}x)$$

Whenever $x = \lambda$ the wave function equals $cos(2\pi)$, whose value is 1, the maximum value of the cosine. Therefore, the standing wave has maximum amplitude every 2π, or every full revolution around a unit circle.

---

[7]There are several simple equations that express the common measurements that apply to waves, and light waves in particular.

wave speed = $\nu \cdot \lambda$, the wave frequency times the wavelength.

In the case of light, the wave speed is a constant, c, so $c = \nu \cdot \lambda$.

Alternately, the wavelength of light is $\lambda = c/\nu$

The angular frequency of a wave, ω is $2\pi\nu$ or the number of oscillations that would occur in a unit circle per second.

The wave number, k is the number of waves that pass in a unit distance. By convention, k is expressed as the number of waves that would pass a point in a circle of unit radius. Accordingly, $k = 2\pi/\lambda$ or $k = 2\pi\nu/c$ or $k = 2\pi/c \cdot \omega/2\pi$ or $k = \omega/c$.

The period of a wave is the time required for one wavelength to pass a point, or the reciprocal of the frequency. Thus, the period of a light wave is $1/\nu = \lambda/c$.

In the case of light waves, the energy is directly related to frequency, wavelength, or period. E = $h\nu$, where h is the Planck constant. Energy in terms of wavelength is $E = hc/\lambda$.

[8]Sharp-eyed readers may have noticed that the equation for the energy of light is dependent only upon the speed of light, and the frequency of the light waves, but not on the amplitude of the light waves at a given frequency. Technically, the equation for the energy of light represents the minimum energy needed for light to propagate with a frequency ν. There is no commonly used method to directly measure the amplitude of light. Instead, physicists use their understanding that light is an electromagnetic wave with the electrical and magnetic components traveling in phase at right angles from each other. Physicists have ways of measuring the electric field strength or the magnetic field strength of light, and such measurements serve to quantify the amplitude of a light wave.

This is how we can write a wave function for a standing wave. In reality, most waves don't stand still, and we need a general wave equation to describe moving waves.

Let's modify our wave function a bit so that our waves can travel. In addition, we will give our new wave function a dignified name, ψ.

$$\psi = e^{i(kx-\omega t)}$$

Here, t is a time interval, and ω is the angular velocity of the wave, and is equal to $2\pi\nu$ where ν is the wave frequency expressed as oscillations per second.

How does this new wave equation help us? Let's consider the case when $kx - \omega t = 0$

In this case, the wave is at a peak amplitude as $\psi = cos(kx - \omega t) = cos(0) = 1$, or $kx = \omega t$ or $x/t = \omega/k$.

But x/t expresses the distance over time or the velocity of the wave. And so, our new formula endows our wave with a velocity.

The new equation also provides a way of understanding wave velocity in terms of wave frequency and wavelength.

$$\text{velocity} = x/t = \omega/k = \frac{2\pi\nu}{2\pi/\lambda} = \nu\lambda$$

It is easy to see now that the wave velocity is the number of waves in a second, (ν), times the wavelength, (λ).

Now that we have used Euler's formula to develop general equations for a traveling wave, we can attach what we know about the physical properties of waves and derive a wave equation that describes the universe as a general wave function. We'll do this in the section titled "Proving Schrödinger's wave equation".

2.7 Proving rotations in the complex plane can be expressed using Euler's formula

Using the example from the prior proof, suppose we rotated z in the complex plane by some angle beta.

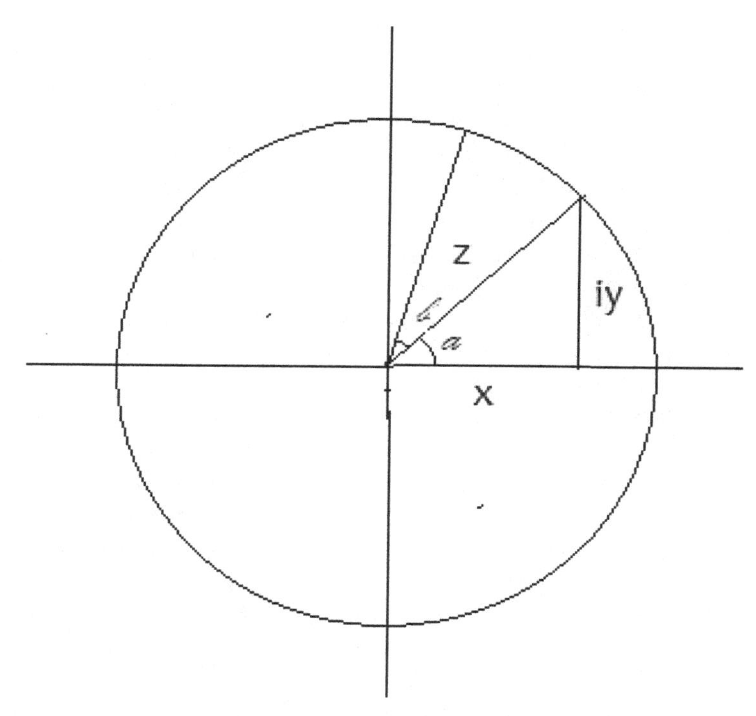

FIGURE 2.3: The complex number at angle alpha has now been rotated by the angle beta, positioning the hypotenuse, z, at a new angle, alpha plus beta.

We have previously shown that z, at angle alpha could be represented by Euler's formula as:

$$ze^{i\alpha} = zcos\alpha + zi\,sin\alpha$$

When we rotate by an additional angle, beta, we see that:

$$ze^{i(\alpha+\beta)} = zcos(\alpha+\beta) + zi\,sin(\alpha+\beta)$$

But we also know:

$$ze^{i(\alpha+\beta)} = ze^{i\alpha}e^{i\beta} = zcos(\alpha+\beta) + zi\,sin(\alpha+\beta)$$

Therefore, the rotation of any complex numbers is equivalent to multiplying the exponential terms of the wave at alpha and the wave at beta. Beta can be considered a change in the phase of the wave. By examining the equation above, it is obvious that it is much easier to calculate rotations (i.e., phase changes) using exponentials than using trigonometric formulas.

2.8 Proving Euler's formula expresses superpositions of waves

All linear systems have a feature known as superpositioning, in which the results of combining system components are accounted for by simply adding one component to the other. Electromagnetic waves have the feature of superpositioning, as do all massless particles. The superpositioning property allows us to tease out individual signals and waveforms from mixtures. For now, let's take a look at how Euler's formula can be used to represent the outcomes of combining wave functions.[9]

$$e^{i\theta} = cos(\theta) + isin(\theta) \qquad \text{Equation 1}$$

Let's substitute $-\theta$ for θ (the formula works for ANY value)

$$e^{-i\theta} = cos(-\theta) + isin(-\theta)$$

We know that $cos(-\theta) = cos(\theta)$ and that $sin(-\theta) = -sin(\theta)$, so: [10]

$$e^{-i\theta} = cos(\theta) - isin(\theta) \qquad \text{Equation 2}$$

Adding equations 1 and 2, we find:

$$e^{i\theta} + e^{-i\theta} = 2cos(\theta)$$

Or,

$$cos(\theta) = (e^{i\theta} + e^{-i\theta})/2$$

Similarly, by subtracting equation 2 from equation 1, we find:

$$sin(\theta) = (e^{i\theta} - e^{-i\theta})/2i$$

So, any combination of sines and cosines can be expressed in terms of Euler's formula. Superpositions of sines and cosines are expressed as additions of sinusoidals (i.e., sums of sines and cosines). Likewise, waves are expressed as sinusoidals. Therefore all superpositions of waves can be expressed in terms of Euler's formula.

[9]We'll be discussing superpositioning in more detail in the section titled "Proving the impulse response convolution theorem". Euler's formula is discussed in the section titled "Proving Euler's formula and Euler's identity".

[10]Cosine and sine are examples of even and odd functions, discussed in section titled "Proving even and odd decomposition applies to all functions".

2.9 Proving fundamental particles and wave packets are equivalent

"There are no facts, only interpretations".
—Friedrich Wilhelm Nietzsche (1844-1900)

On a simplistic level, we can prove the equivalence of fundamental particles and wave packets by specifying the definitions of each, and recognizing that the definitions are equivalent.

- Let's define what we mean by a "wave packet". A wave packet is a wave composed of a collection of superpositioned waves whose frequencies cluster around a single value. For example, the wave packet known as an electron is a wave whose composite wave frequencies are tightly clustered around a value that is the characteristic frequency of the electron. The electron wave-packet behaves in a manner that is characteristic of its role as the quantum of the electromagnetic field.
- A physical particle is a set of field properties at a point in space. As such, it has no size, and its energy is determined by the mechanics of its motion. Because fundamental particles have no size, we cannot "see" an individual particle. We can only detect the presence of a particle due to its interaction with physical systems (i.e., sensors). Our functional definition of a particle is that it is whatever is detected by a sensor tuned to its energy.
- A physical detector cannot distinguish a wave packet (characteristic energy at a location) from a particle (characteristic energy at a location).
- There being no physical way to distinguish a wave packet from a particle, then the two must be equivalent.

Of course, our "proof" does not tell us very much about the nature of wave packets or of particles. We really need to step back and ask ourselves what it means to be a particle composed as a wave packet.

A single wave has a pure sinusoidal pattern. What happens when we superposition multiple waves? As we add sinusoidals together, we get a combined wave envelope. Sound engineers refer to the combined wave envelope as the "beat". When many combined waves, all with nearly the same frequency, are combined, they yield a single prominent moving peak, surrounded by small ripples at all other locations. The attenuation of the combined wave outside of its sharp peak is due to the inability of many waves to stay in phase outside of the point where the peak occurs. Once out of phase, they interfere with one another, canceling their combined amplitude. Beyond the peak, the likelihood that the component waves will get back into the same phase again (forming additional peaks) drops off to something very small. The single-moving peak is known as the wave packet.

The wave packet can be detected by instruments designed to register point-like pulses of energy, so-called particle detectors. To the detector, a particle is any pulse of energy that can be detected by the sensor and assigned a particular set of coordinates (i.e., the location in the sensor where the detection was registered). We cannot see fundamental particles, so we depend upon our particle detectors to tell us when and where a particle has triggered its sensor. For all purposes, the detector creates something that we interpret as a particle (i.e., a quantal object with the energy to register on the sensor, at a determined location). **Hence, as detected by the sensor, there is no observable distinction between a wave packet and a particle.**

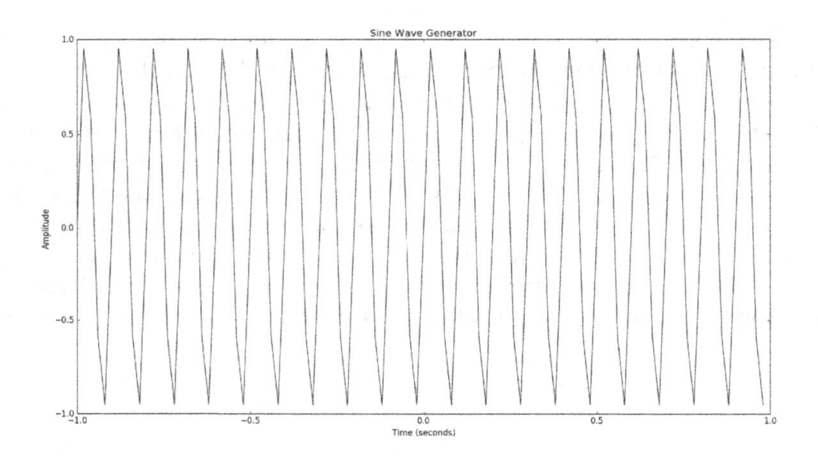

FIGURE 2.4: A sine wave, oscillating at 10 cycles per second with a peak amplitude of 1.0.

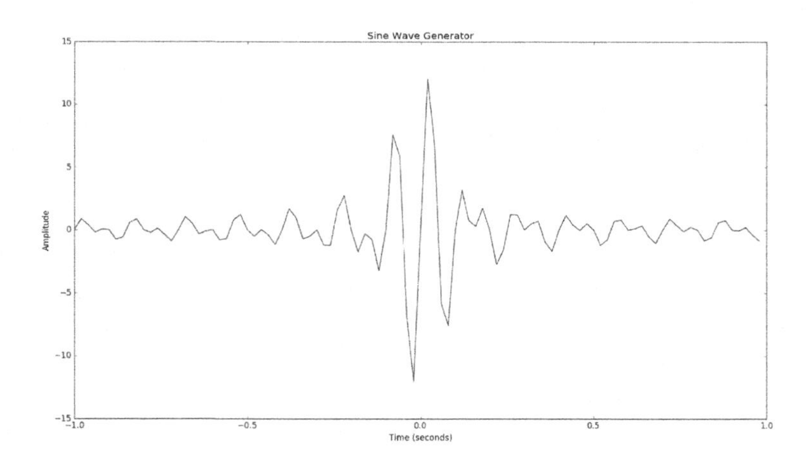

FIGURE 2.5: Something akin to a wave packet formed from adding together 14 sinusoidal waves to the wave shown in the figure above, all with nearly the same frequency and amplitude. There is one central peak formed (i.e., the wave packet), with only small waves everywhere else.

2.10 Proving wave packets move more slowly than their constituent waves

Here's the problem:

1. If particles are composed of superpositioned waves of their respective fields (i.e., as wave packets),

   And,

2. If the waves of any field are massless objects that must move at the speed of light,

   Then,

3. All fundamental particles must move at the speed of light.

   But,

4. Electrons are fundamental particles that have mass and that cannot move at the speed of light.

   So,

5. How can our original assertion - that fundamental particles are wave packets—be true?

It turns out that there is a very simple solution to this puzzle, and the relevant physical principles were developed by William Rowan Hamilton (1805-1865), in 1839. He found that when a wave is composed of the superposition of multiple waves, the velocity of the composite wave (the group velocity) is different from the velocity of the individual waves (the phase velocity). Because wave packets are composite waves, we can infer that wave packets can move more slowly than the speed of light. But let's not take Hamilton's word for it. Let's prove for ourselves that combining waves may result in a wave packet that moves slower than the speed of light. While we're at it, we'll show how motionless wave packets can be constructed from sinusoidal waves, and that a wave packet may move in the opposite direction of its composite waves.

First, let's take a look at what happens when we sum two sinusoidals (i.e., when we superposition two sinusoidals) having different frequencies and phases. The result will be a composite sinusoidal wave. For example, we can sum $\sin(x) + \cos(0.9x)$ to produce a wave that seems to envelop smaller waves, much like a large ocean wave may envelop the swell produced by contributing small ripples. We call this enveloping wave the "beat wave", and we can prove that its speed is less than the speed of light.

Our proof involves nothing more than summing the general wave equations for two waves and evaluating the result. Our findings for the summation of two waves will also apply to the summation of any number of waves, because the sum of sinusoidals always yields another sinusoidal.

To begin, let's review two trigonometric equations that will come in handy in a moment:

$$\sin(\alpha) + \sin(\beta) = 2\sin(\frac{\alpha+\beta}{2}) \times \cos(\frac{\alpha-\beta}{2}) \quad (1)$$

and

$$\cos(\alpha) + \cos(\beta) = 2\cos(\frac{\alpha+\beta}{2}) \times \cos(\frac{\alpha-\beta}{2}) \quad (2)$$

Now, let's look at the wave equations of two arbitrary sinusoidals, Ψ_1 and Ψ_2:

$$\Psi_1 = cos(\alpha) + isin(\alpha) = e^{i\alpha}$$

and

$$\Psi_2 = cos(\beta) + isin(\beta) = e^{i\beta}$$

Just for now, we do not need to know the values of α or β. Let's add the two wave equations to produce a new wave equation, which we'll call Ψ_{combo}.

$$\begin{aligned}&\Psi_{combo}\\&= \Psi_1 + \Psi_2 = cos(\alpha) + isin(\alpha) + cos(\beta) + isin(\beta)\\&= cos(\alpha) + cos(\beta) + i(sin(\alpha) + sin(\beta))\end{aligned}$$

From our two trigonometric equations,

$$\begin{aligned}&\Psi_{combo}\\&= 2\cos(\frac{\alpha+\beta}{2}) \times \cos(\frac{\alpha-\beta}{2}) + i\Big(2\sin(\frac{\alpha+\beta}{2}) \times \cos(\frac{\alpha-\beta}{2})\Big)\\&= 2\cos(\frac{\alpha-\beta}{2})\Big(\cos(\frac{\alpha+\beta}{2}) + i\sin(\frac{\alpha+\beta}{2})\Big)\\&= 2\cos(\frac{\alpha-\beta}{2})e^{i(\frac{\alpha+\beta}{2})}\end{aligned}$$

Believe it or not, we're done! All that's left for us to do is to comprehend the significance of our result. We see that the combined wave function, Ψ_{combo} is equal to a new wave function, $e^{i(\alpha+\beta)/2}$ multiplied by a sinusoidal harmonic, $2\cos((\alpha-\beta)/2)$.[11] Whenever α equals β, the value of the harmonic

[11] Remember that the cosine of an angle is equivalent to a sine of the same angle shifted by $\pi/2$ radians. Consequently, cosinusoidal oscillations are equivalent to shifted sinusoidal oscillations.

is 1, and the value of the combined wave is at its peak (i.e., the sum of the two contributing waves). When α and β differ by a factor of $\pi/2$, the value of the sinusoidal drops to zero, and the value of the combined wave drops to zero. The sinusoidal modulates the value of the combined wave function, Ψ_{combo}, and produces a combined wave pattern that looks something like the enveloped wave is shown in our figure. That is to say that there is a "group" wave covering the small oscillations of its contributing small waves.

How does the velocity of the group wave compare with the velocities of its constituent superpositioning waves (that move with the speed of light)? To answer this question, we must return to the wave formula that describes α and β.

Our general wave formula is:

$$\Psi = e^{i(kx+\omega t)}$$

As discussed in the section titled "Proving that waves can be expressed in terms of Euler's identity", the wave number, k, is the number of waves that pass in a unit distance, and ω is the angular frequency of the wave. The velocity of any wave is $\frac{k}{\omega}$.

We can restate the wave functions for all of our waves:

$$\begin{aligned} &\Psi_1 = e^{i\alpha} = e^{ik_1x+\omega_1} \\ &\Psi_2 = e^{i\beta} = e^{ik_2x+\omega_2} \qquad \text{where} \\ &\alpha = k_1x + \omega_1 \qquad \text{and} \\ &\beta = k_2x + \omega_2 \end{aligned}$$

As previously noted, the group sinusoidal for the the superposed wave, Ψ_{combo} is

$$2\cos(\frac{\alpha-\beta}{2})$$

And $\frac{\alpha-\beta}{2}$ describes the motion of the oscillation. Evaluating, we find:

$$\begin{aligned} &\frac{\alpha-\beta}{2} \\ &= \frac{k_1x + \omega_1 - k_2x - \omega_2}{2} \\ &= \frac{(k_1 - k2)x + (\omega_1 - \omega_2)}{2} \end{aligned}$$

We know that the velocity of a wave is $\frac{\omega}{k}$ [12] In the case of the group wave, its

[12] Discussed in the section titled "Proving that waves can be expressed in terms of Euler's identity".

velocity is, then:

$$\text{velocity of the group wave} = \frac{\omega_1 - \omega_2}{k_1 - k2}$$

Or, as the limiting difference in the superpositioning waves approaches zero, the group wave velocity equals $d\omega/dk$. By playing with the available variables for our superpositioning waves (i.e., variables ω and k), we can produce a group wave that moves at virtually any speed (not exceeding $\pm$ the speed of light), and moving in the forward or backward direction.

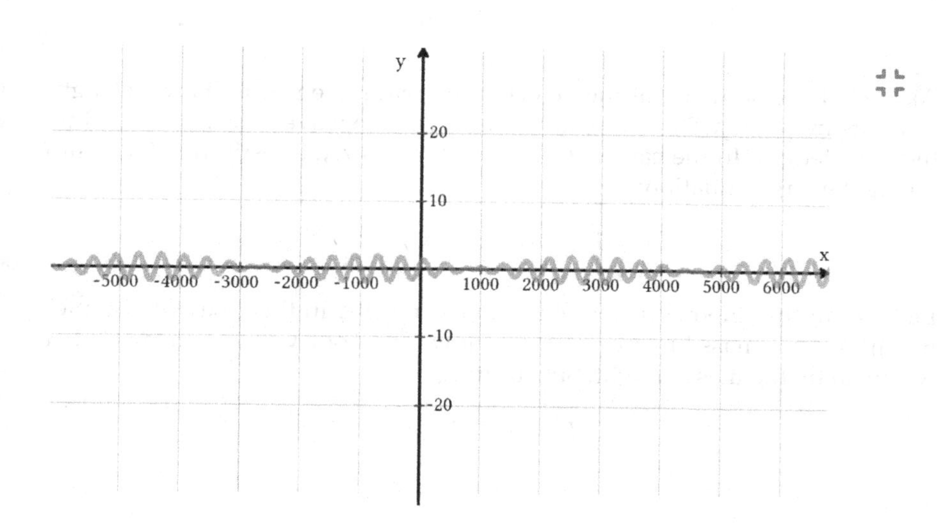

FIGURE 2.6: The wave pattern resulting from adding two sinusoidals, $\sin(x) + \cos(0.9x)$. We can discern a long, large wave composed of many small waves.

2.11 Proving Schrödinger's wave equation

Now that we have learned how to express waves as mathematical equations, all we need to do is to attach the known physical properties of waves to the abstract wave equations to produce the descriptive formula known Schrödinger's wave equation (named for Erwin Schrödinger. To do so, we need to visit Einstein's famous formulas for the energy of light in terms of either mass or frequency, where c is the speed of light, h is the Planck constant, and ν is the frequency of light.

$$E = mc^2 = h\nu$$

Also, we have shown that the velocity of a wave is equal to its wavelength times its frequency. Therefore, its frequency is the wave velocity divided by the wavelength. In the case of light waves, $\nu = c/\lambda$. Substituting back into the light energy equation:

$$E = mc \times c = hc/\lambda$$

Let's recall that momentum = mass times velocity. In the case of light, momentum = p = mass times c = mc. For now, let's not try to understand what we mean by the mass of light. Substituting p for mc,

$$E = pc = hc/\lambda$$

Therefore,

$$\lambda = h/p$$

The wave number, k, is the number of wavelengths in one revolution or cycle. We have previously shown that $k = 2\pi/\lambda$. So, $k = 2\pi p/h$.

By convention, we call $h/2\pi$ by the symbol $\hbar$. So, $k = p/\hbar$.

Now, let's return to the general wave equation for a traveling wave, derived in the section titled "Proving waves can be expressed in terms of Euler's formula".

$$\psi = e^{i(kx-\omega t)}$$

Let's take the derivative to see how the wave function changes over distance.

$$\frac{d\psi}{dx} = ike^{i(kx-\omega t)} = ik\psi$$

The derivative of the derivative (second derivative) yields,

$$\frac{d^2\psi}{dx^2} = (ik)^2\psi = -k^2\psi$$

Substituting $p \div \hbar$ for k and rearranging,

$$-\hbar^2 \times \frac{d^2\psi}{dx^2} = p^2\psi$$

Recalling our high school physics, the total energy in a system is equal to the kinetic energy plus the potential energy, U, or $E = K.E. + U$.

The kinetic energy is equal to one-half the mass times velocity squared, written as: $K.E. = \frac{1}{2}mv^2$.

The momentum of a moving object, p is mass times velocity. We'll represent momentum as p and write our equation: $p = mv$[13, 14]

Let's rewrite the energy equation, remembering that $p = mv$ and $v = p/m$. Substituting back into the original equation,

$$E = \frac{1}{2}mv^2 + U$$

Or,

$$E = \frac{1}{2}\frac{p^2}{m} + U$$

Multiplying both sides of the equation by our wave function,

$$E\psi = \frac{p^2\psi}{2m} + U\psi$$

Remembering that,

$$\psi = -\frac{\hbar^2}{p^2}\frac{d^2\psi}{dx^2}$$

[13]Sharp readers will recognize here that momentum is just the time derivative of the energy.

[14]You will notice that the same relationship between momentum and mass holds for solid matter and for waves, a notion developed in 1924 by Louis de Broglie and reviewed in his book, *The Theory of Measurement in Wave Mechanics (Usual Interpretation and Causal Interpretation)*, Gauthier-Villars, Editeur-Imprimeur-Libraire, 1957. A simple argument supporting de Broglie's theory is that if electrons and other subatomic particles exhibit wave/particle duality, and if all matter is composed of subatomic particles, then we would expect all matter to exhibit wave properties and to conform to the fundamental wave formula expressed in the Schrödinger wave equation.

We get,

$$E\psi = -\frac{\hbar^2}{2m}\frac{d^2\psi}{dx^2} + U\psi$$

This is the time-independent Schrödinger wave equation, and it tells us how waves evolve over distance. Here, we took the derivative of the wave equation with respect to distance, from which we derived our time-independent equation. We could have just as easily taken the derivative of the wave equation with respect to time, and that would have yielded the distance-independent Schrödinger wave equation. We mention here that physicists employ a general form of the Schrödinger wave equation, employing the Hamiltonian operator (upon which we shall not dwell). For the record, it looks like this:

$$i\hbar\frac{\partial}{\partial t}|\psi(t)\rangle = \hat{H}|\psi(t)\rangle$$

The generalized wave equation tells us how the state of any quantum object, such as an electron, will evolve over time.

The purpose of this exercise is to show the reader that one of the most profound formulas in quantum physics, that describe the evolution of all waves and matter—matter being just another form of wave—over time and space, can be derived using simple high school mathematics.

The Schrödinger wave equation provides a handy wave model of quantum physics. There is no proof of the physical validity of the equation, and physicists are keenly aware of its short-comings. These include various quandaries such as the collapse of the wave equation (an apparent breakdown of causality), negative probabilities, and infinite towers of negative energy states.[15]

[15]Discussed in depth in David Tong's *Quantum Field Theory: Part III Mathematical Tripos*, University of Cambridge, Michaelmas Term, 2006 and 2007.

2.12 Proving the probability density for the wave equation is $|\psi|^2$

Let's take a look at a very simple example of a stationary wave, described in the following equation,

$$\psi = e^{ikx} = e^{\theta} = \cos\theta + i\sin\theta$$

Now, let's evaluate the modulus square of psi, defined as,

$$|\psi|^2 = \psi \times \psi^* = (\cos\theta + i\sin\theta)(\cos\theta - i\sin\theta)$$

Expanding the term on the right,

$$|\psi|^2 = (\cos\theta)^2 + (\sin\theta)^2 = 1$$

So, the square modulus of the simplest form of a wave equation equals 1, and this fulfills a normalizing condition for a wave equation describing a wave that must exist somewhere. That is to say when we add up all the probabilities of a wave being at all possible locations, the probability of finding it in one of those locations must be 1.

Wait a minute! Is this some sort of bait-and-switch operation? When we defined the wave equation, we were simply coming up with a function that described the progress of the wave over time. Probabilities were not considered. Why would the value of $|\psi|^2$ have anything to do with wave probabilities?

The standard answer to this question is circular in nature, and not particularly convincing. We know that $|\psi|^2$ represents the length of a wave vector. This is simply the Pythagorean theorem applied to the real and imaginary components of the wave. We also know that any wave function could be represented as the summation of any number of component wave functions, each of which could be represented by its own wave functions, all having smaller vectors. The square modulus of each of those component wave functions would need to add up to a grand total of 1 to fulfill the normalization condition (i.e., the probability that the combined wave function is located somewhere in the universe). We can also imagine that the magnitude of the square modulus of any individual components must be proportional to the probability of the aggregate wave occupying its position. In sum, the square modulus of the wave function, $|\psi|^2 = \psi \times \psi^*$ fits the criteria that we need to find in a wave probability density; and that is why we use it.

It is prudent not to become overly enthralled with probability representations for the wave function. The wave function itself is a useful but somewhat makeshift model of reality. Probability functions based on the wave function are convenient and practical extensions of the wave model that do not lend themselves to formal proof.

2.13 Proving the speed of light is constant

Just for fun, let's imagine a universe wherein light of different wavelengths traveled at different speeds. For example, blue light might travel at a speed 10% greater than the speed of red light. If this were the case, white light, which is composed of all the different colors (i.e., wavelengths) of visible light, would quickly lose its coherence and it's "whiteness". Emitted white light would reach us as blue light first, and the red light would come in some time later. We never observe this kind of behavior in our light.

Let's imagine another scenario entirely, in which the speed of light depended upon the location at which the light was emitted. We might have a situation in which a pulse of light leaves a star at a distance of 1 billion miles from Earth. Along the way, half of the light is absorbed and re-emitted by a mirror located in a region of space where emitted light travels at 10 times the speed of light. In this case, we would see the pulse from the star arriving at one time (from the fast-moving re-emitted source) and the slower light from the same pulse would arrive sometime later. Again, such occurrences are at odds with reality.[16]

Actually, there is a very simple way to prove that the speed of light must be constant. Everyone knows Einstein's relationship between the rest mass of an object and the speed of light.

$$E = mc^2$$

Suppose the speed of light were not constant, changing from moment to moment. For example, suppose the speed of light suddenly doubled. Then we would have,

$$E = m(2c)^2 = 4mc^2$$

When the speed of light doubles, the associated energy increases 4-fold, without adding any energy to the system. This is a clear violation of the law of conservation of energy. Therefore, the speed of light must be constant. The constancy of the speed of light was demonstrated experimentally in the Michelson-Morley experiment, discussed in some detail in the section titled "Proving that there is no ether."

[16]Readers may object here that gravity lensing, a phenomenon that occurs when light from a distant star passes through a high gravity area of space will arrive on Earth at different moments of time, and will seem to come from different points in space. This phenomenon is due to gravity's distortion of space (sometimes referred to as the bending of light by gravity), causing portions of the emitted light to follow different paths. In this instance, the speed of light is not changed, but the angle of the path, and its distance to Earth, are subtly altered.

Let's look at another proof of the constancy of the speed of light, based on a simple gedanken experiment, credited to William de Sitter. Imagine that we have our telescope trained on a luminous body rotating around a point in distant space. We observe it going around and around in its orbit, as a single point of light tracing a smooth ellipse. Now imagine that the speed of the light emanating from the body is variable. One moment it travels at the traditional speed of light; the next moment it travels a bit faster, and at other moments it travels a bit slower. In that case, the light received at our telescope will consist, at any given moment, of a combination of light beams originating at different moments, at different positions in the orbital ellipse, but "catching up" to one another when they reach the telescope. What we would see might be multiple luminous bodies, spaced at different locations in the elliptic, or we might have moments when our image goes totally blank because the light from the body has arrived previously or has yet to arrive. As we observe, the continuous reception of a single point of light tracing a continuous elliptic curve informs us that the light emitted from the orbiting body must have traveled at a constant velocity.

2.14 Proving photons are massless

For this proof, we need to accept that the total energy of any object (e.g., a fundamental particle, mosquito, a baseball) is its rest energy and it's energy of motion, combined. Rest energy is the energy of a motionless object and is equal to the rest mass of the object times the speed of light squared.

$$E_{rest} = m_{rest}c^2$$

Now, let's see what we can infer about the energy associated with light waves.

1. Light (a moving photonic wave) has no frame of reference in which it is motionless.

   Because,

2. Light always travels at the same speed (i.e., the speed of light) when observed in every frame of reference.

   Therefore,

3. Light is never at rest, and consequently has no rest mass.

   But,

4. If light has no rest mass, then all of its energy comes from its motion.

   Therefore,

5. There is no mass associated with the energy of light.

   But,

6. Photons are the quanta of light in motion.

   Therefore,

7. Photons have no mass.

If photons have no mass, then what accounts for the motion energy of light? Although photons have no mass, they do have momentum and the momentum of light accounts for the energy of light waves. We'll be returning to the momentum of photons in further sections (when we discuss particle spin). For now, keep in mind that our proof applies generally. **Any particle that moves at the speed of light must be massless.**

2.15 Proving massless light carries energy

"Nothing happens until something moves".
—Albert Einstein

In the previous section, we showed that particles that move at the speed of light are massless. But hasn't Einstein proven that energy and mass are interchangeable? In that case, if light (composed of photons) has no mass, then light should have no energy. Does this make any sense?

Actually, no. Light has lots of energy. In point of fact, most of the energy that we can observe, on a cosmologic scale, comes from electromagnetic waves that were produced by the Big Bang and which pervade all of spacetime as the cosmic background radiation (see section titled "Proving most of the observable energy of the universe is the cosmic microwave background"). A smaller fraction of the energy in our universe comes from the light emitted by stars (e.g., the sunlight that heats our atmosphere). Other sources of energetic electromagnetic waves come from the radioactive decay of heavy elements and from photon emissions from atomic electrons as they transition to lower orbitals.

The energy of light must come from somewhere, but it does not come from its rest mass, because photons are massless. As it happens, all of the energy of light comes from the energy produced by its motion, and light's energy of motion comes entirely from its momentum, p.

$$E^2 = (pc)^2$$

In introductory physics courses, were are taught that momentum is equal to mass times velocity. In the case of light, there is no mass (previously proven), so we might guess that its momentum would equal zero and that the energy of light could not possibly come from its momentum. We would be wrong. How does light, having no mass, acquire momentum? The momentum of light comes from its spin. Spin is an intrinsic property associated with all waves. All matter can be described by waves, so all of the particles that constitute matter must have an assigned spin. But where does the spin come from? In the case of electromagnetic radiation, spin momentum comes from the magnetic component of the wave. Oscillations in the magnetic field produce an additive vector force with each peak and trough of its oscillation. The oscillations in the electric field produce a vector that cancels out with each oscillation cycle and do not contribute to the momentum of light.

2.16 Proving massless particles cannot move slower than light

"Vita in motu.
(Life is in motion.)"
—Latin motto

The currently recognized massless particles are photons and gluons.[17] If we wanted to be cute, we could argue that photons must always move at the speed of light (and no slower than the speed of light) because photons are light, and light always moves at the speed of light. We can construct a more general proof to cover all massless particles (known or unknown), as follows:

1. Imagine that we had a massless particle that moves slower than the speed of light.
2. If the particle is moving slower than the speed of light, then there is a frame of reference in which the particle is motionless

   This is equivalent to saying that an observer could match its speed, at which point the observed velocity of the particle would be zero.
3. Furthermore, all frames of reference are equally valid.

   That is to say that measurements of the massless particle are valid within its zero-velocity reference frame.
4. In the zero-velocity reference frame, the massless particle would have no energy.

   Energy is what makes particles move. If the velocity of the particle is zero, and there is no movement, and there is no mass, then there is no energy.
5. A particle with no size, no mass, no movement, and no energy is a non-existent particle.[18]

   Therefore,
6. Our initial premise is impossible, and there can be no frame of reference in which a moving observer can match the speed of a massless particle.

[17] In addition, one type of neutrino may be massless, and the as yet unconfirmed graviton particle may also be massless.

[18] This calls to mind one of my favorite limericks: "As I was sitting in my chair, I saw the bottom was not there; nor legs nor back; but I just sat; ignoring little things like that."

7. The speed that cannot be matched by a moving observer is the speed of light.

   Therefore,

8. **All massless particles move at the speed of light (at the very least).**

2.17 Proving massless particles cannot move faster than light

Here are three short arguments against faster-than-light speed for massless particles.[19]

- Argument 1. As a particle moves faster and faster, its clock slows down (as per special relativity). When it reaches the speed of light, time stops completely and the particle exists in the so-called "eternal now".[20] It is meaningless to imagine a particle moving at a speed greater than the speed of light because once time has stopped, there is no way in which time can be further stopped, and greater-than-light speed becomes meaningless.
- Argument 2. If a massless particle were to exceed the speed of light, then it would become unobservable (by outpacing the light by which they could be observed). An unobservable object, with no mass and no size (fundamental particles have no size) cannot arguably exist (there is nothing there, really). Therefore, if a particle were to move faster than the speed of light, it would exit existence.[21]
- Argument 3. Objects moving faster than the speed of light would violate the dictum that observations are valid in all reference frames. Observations would not be repeatable in different reference frames because the light that captures an event could never reach the observer traveling at greater than light speed.

[19] Every high school student is taught that massive particles cannot move at the speed of light the energy required to move a mass varies as the inverse of $\sqrt{1 - v2/c2}$. As $v \rightarrow c$, the energy approaches infinity. Therefore, no finite amount of energy can move a mass at the speed of light or greater.

[20] The term "eternal now" seems to have made its first appearance in Karl Pearson's *The Grammar of Science*, published in 1900 by Adam and Black. Pearson's book was studied by Albert Einstein, and it is likely that the notion of the "eternal now" played some role in Einstein's development of the theory of special relativity.

[21] This suggests yet another reason why massive objects cannot exceed the speed of light. A massive object is composed, in part, by massless fundamental particles (quarks, gluons, and photons). An aggregate object cannot move faster than the fastest moving particle of which it is composed.

2.18 Proving the Lorentz transformation

The famous Lorentz transformation of special relativity is an equation that describes a spacetime invariance. The invariance is that the observed interval between any two events is the same regardless of the relative motion of observers (in different frames of reference). For example, one observer may be traveling at 1 mile per hour, and another observer may be traveling at 1 million mile per hour, but they will both measure the same interval separating two events as observed from their own reference frames. The reason for the invariance is that space is homogeneous (the same everywhere, with all of the space exhibiting the same physical laws) and isotropic (all directions around a point are equivalent and measurements made at a distance from the point will be the same, in all directions).

Using Euclidean spacetime coordinates,

$$c^2(t_2 - t_1)^2 - (x_2 - x_1)^2 - (y_2 - y_1)^2 - (z_2 - z_1)^2$$
$$= c^2(t'_2 - t'_1)^2 - (x'_2 - x'_1)^2 - (y'_2 - y'_1)^2 - (z'_2 - z'_1)^2$$

Here we use x, y, z as the typical 3-D distance coordinates, The time coordinate, expressed as distance, is ct. The above expression simply expresses the invariance, recognizing that distances between the origin (0,0,0,0) and a point in spacetime is $x^2 + y^2 + z^2 - c^2t^2$.

At its most basic level, this is all that the Lorentz transformation says, but it serves as the geometric embodiment of all of special relativity.[22, 23] The Lorentz group accounts for special relativity insofar as special relativity represents the spacetime invariance of distance expressed as $x^2 + y^2 + z^2 - c^2t^2$ and can be considered equivalent to the group of rotations in spacetime (i.e., the 4 coordinates x, y, z, and t).[24]

What is the significance of the "negative" sign in the Lorentz transformation? Specifically, why is the time coordinate (expressed as "ct") always the opposite of the sign of the x, y, and z coordinates? In spacetime, wouldn't

---

[22]The Lorentz group is the group of all Lorentz transformations of Minkowski spacetime.

[23]Named for Hendrik Lorentz (1853-1928) and Hermann Minkowski (1864-1909).

[24]In later sections we will view the Lorentz transformations as the substitutions of the variables x,y,z,t that leave the quadratic form of $x^2 + y^2 + z^2 - c^2t^2$ invariant. These transformations constitute a symmetry group. For the definition of a group, see section titled "Proving the uniqueness of inverse elements of groups". Group theorists recognize the Lorentz group as O(1,3), the indefinite orthogonal group of linear transformations of n-dimensional space. Doing so permits mathematicians to treat the group as matrices having a particular set of properties that lend themselves to formulaic analysis.

we expect that all coordinate distances in a homogeneous and isotropic universe would have the same sign? Furthermore, wouldn't we expect that the Pythagorean distances, calculated from Euclidean coordinates, would equal $x^2 + y^2 + z^2 + c^2t^2$, and not $x^2 + y^2 + z^2 - c^2t^2$?

We can see why the time coordinate is negative by imagining the following situation. You are gazing at a star that is 10 light years distant from Earth. What is the total spacetime distance from your eye to what you see in the telescope? The answer is zero centimeters. The reason for this surprising answer is that when you look at an image in a telescope, you are not seeing the star in its current spacetime position. You are seeing the light that has traveled from the star to your eye, a voyage that took, in this case, 10 years. The image you see is a record of the past, not of the present. It would be absolutely impossible to see the star in its current position, insofar as doing so would violate causality. The best we can ever do is to wait for the light from the star to reach our eyes, at which point, its distance from us is zero. How would we calculate the distance to the light's origin? That would be the Pythagorean distance calculated for Euclidean coordinates (i.e., $x^2 + y^2 + z^2$ minus the square of the distance traveled by the light emitted by the star (i.e., the distance light travels in 10 years). The minus sign is present in the Lorentz transformation because we can only see things that happened in the past. We cannot see the present or the future. Distance in time, as we observe it, is always negative.[25]

[25]This topic is explored in greater detail in the section titled "Proving that only the past is observable".

2.19 Proving special relativity by Doppler

"You can observe a lot by watching".
—Yogi Berra

All of special relativity stems from two fundamental conditions of spacetime:

- The laws of physics are the same in all inertial frames of reference.
- The velocity of light is the same for all inertial observers.

The second condition was shown to be true experimentally by Michelson and Morley, in 1886 (see discussion in section titled "Proving that there is no ether"), but anyone could have inferred the second condition from the first, with a little thought.

1. If the laws of physics are the same in all inertial frames of reference, then
2. The observables must be the same in all inertial frames of reference. And,
3. What we observe is dependent upon the speed of light.
4. Therefore, the speed of light must be the same for all inertial observers.

So, knowing only that the laws of physics are the same in all inertial frames of reference, scientists had all they needed to develop the principles of special relativity.[26] This being the case, it is something of a mystery as to why the world needed to wait until 1905 for Albert Einstein to enlighten us.

In point of fact, there are several ways in which the special relativity could have been derived, without resorting to modern advances in physics and mathematics. One such way depends on nothing more than an understanding of the Doppler effect, discovered in 1842 by Christian Doppler. Here, we

[26]Note on terminology. What we call special relativity is a misnomer, as it suggests a brand of relativity that applies only in special circumstances; while classical relativity (the relativity of Galileo Galilei and of Isaac Newton) would apply under normal circumstances. Of course, special relativity applies in all circumstances, and classical relativity applies only when dealing with objects in slow motion; and then only approximately. When we think about it, everyday motions of atoms, electrons, planets, stars, and galaxies all occur at so-called relativistic speeds. Clearly, the relativity of Newton and Galileo is special, and the relativity of Einstein is general. Unfortunately, the "general" term, as applied to relativity, is reserved for gravity, so we are stuck with a less than accurate terminology.

shall derive the time dilation formula of special relativity, using Christian Doppler's mid-nineteenth century observation.[27]

We'll begin with a light source moving at a constant velocity toward one observer and away from a second observer.

Let's look at our variables:

- ObserverToward = the observer to which the light is approaching.
- ObserverFrom = the observer from which the light is receding.
- v is the velocity of the light source.
- ω is the frequency of light.
- ω_0 is the frequency of the emitted light when stationary.
- ω_v is the frequency of the emitted light moving at velocity, v.
- λ is the wavelength of light.

We will soon need to evaluate the ratio of ω_v/ω_0 and although we do not know what this equals, we do know that the ratio will be a dimensionless number. Therefore, we can infer that the ratio can be represented by a function of v/c, a dimensionless number computed from the velocity of the emitter and the velocity of light. That is to say: $\omega_v = \omega_0 \times g(\beta)$ where $\beta = v/c$ and $g(\beta)$ is some function of β.

Δt is the time between two pulses from the light source. Note that since the observers are receiving two pulses of light, the time between pulse 1 and pulse 2 represents the time for light to travel the distance between the two pulses, which the observers would perceive as one wavelength.

As perceived by ObserverToward, the Doppler effect compresses the wavelength, which in this case is the distance measured between the first pulse and the second pulse. The wavelength measured by ObserverToward is:

$$c\Delta t - v\Delta t = (c - v)\Delta t = \lambda_{ObserverToward}$$

The wavelength perceived by ObserverFrom will be broadened as the light source recedes, as:

$$c\Delta t + v\Delta t = (c + v)\Delta t = \lambda_{ObserverFrom}$$

[27]Marco Moriconi derived the special relativity formulas for time dilation, length contraction, addition of velocities, and the mass-energy relation using nothing more than his understanding of the Doppler effect. In this section, we generally follow Moriconi's work, "Special Theory of Relativity through the Doppler Effect" in: arXiv:physics/06052040v1, May 24, 2006.

The general formula for the speed of any wave is:

wavelength times the frequency of the wave = wave speed

Now that we have assembled all of our variables, plus the general formula for the speed of a wave, plus our knowledge that both observers must find that the measured speed of light must equal the constant, c, we can compare the findings of the two observers (ObserverToward and ObserverFrom).

ObserverToward finds that the speed of light is

$$\lambda_{ObserverToward} \times \omega_v = ((c - v)\Delta t)\omega_v = c$$

In the case of ObserverFrom, the frequency of the light is the frequency occurring at $-v$ because the wave is moving away from the observer. ObserverFrom finds that the speed of light is

$$\lambda_{ObserverToward} \times \omega_{-v} = ((c + v)\Delta t)\omega_{-v} = c$$

Dividing the first equation by the second and making our substitutions, (i.e., $g(\beta) = \omega_v / \omega_0$ and $\beta = v/c$. we find:

$$\frac{g(\beta)}{g(-\beta)} = \frac{1+\beta}{1-\beta}$$

The next part of the derivation is a bit tricky, requiring us to imagine an observer moving toward the emitter at speed v (this being equivalent to the emitter moving toward the observer at speed v).

The equation for the frequency of light observed by the observer at rest is

$$\omega_v = \omega_0 \times g(\beta)$$

The emitter "looking" at the observer at rest will receive light at a frequency of $\omega_v \times g(-\beta)$ or $\omega_0 \times g(\beta) \times g(-\beta)$.

But both the observer and the emitter are in the same rest frame and will obviously both observe light at the same frequency, v_0.

So,

$$\omega_0 = \omega_0 \times g(\beta) \times g(-\beta)$$

Therefore,

$$1 = g(\beta) \times g(-\beta)$$

Returning to our prior equation:

$$\frac{g(\beta)}{g(-\beta)} \times (g(\beta) \times g(-\beta)) = \frac{1+\beta}{1-\beta}$$

Or,

$$g(\beta)^2 = \frac{1+\beta}{1-\beta}$$

And,

$$g(\beta) = \left(\frac{1+\beta}{1-\beta}\right)^{1/2}$$

Plugging our derived value for $g(\beta)$ back into our relationship between doppler frequency and emitter frequency, **we have the formula for the relativistic Doppler effect.**

$$\omega_v = \omega_0 \times g(\beta) = \omega_0 \times \left(\frac{1+\beta}{1-\beta}\right)^{1/2}$$

Now, we can proceed to derive the relativistic time dilation formula. Let's go back to our original Gedanken experiment, in which a pulse of light is emitted by a light source moving toward an observer (ObserverToward) at speed v. Remember, between these two pulses, one wave is emitted and received by the observer. The formula for the speed of light, as perceived by ObserverToward, must equal the constant speed of light (our second postulate).

$$\lambda_{ObserverToward} \times \omega_v = ((c-v)\Delta t)\omega_0 \left(\frac{1+\beta}{1-\beta}\right)^{1/2} = c$$

Using the same formulaic approach, we can compute the perceived speed of light from the light-source frame of reference, in which $v = 0$, and $\beta = v/c = 0$ and $\Delta\tau$ is the perceived time interval between pulses.

$$c\Delta\tau\omega_0 = c$$

So,

$$(c-v)\Delta t\omega_0 \left(\frac{1+\beta}{1-\beta}\right)^{1/2} = c\Delta\tau\omega_0$$

Or,

$$\Delta t = \frac{\Delta \tau}{(1 - \beta^2)^{1/2}}$$

Substituting back $\beta = v/c$, **we obtain the relativistic time dilation formula.**

$$\Delta t = \frac{\Delta \tau}{\sqrt{1 - v^2/c^2}}$$

2.20 Proving $E = mc^2$

Once we have the relativistic Doppler effect formula (see section titled "Proving special relativity by Doppler"), we can easily prove the $E = mc^2$.[28]

Imagine the following experiment, wherein we are traveling in the frame of reference of a rocket ship moving at velocity v and watching a cat in a stationary reference frame (relative to our ship). To us, the cat is moving at velocity v, with a kinetic energy of $\frac{1}{2}m_{beforeflash}v^2$ Suddenly the cat emits a flash of light. By doing so, the cat instantly becomes lighter, having lost the mass associated with the energy of the emitted light. Because we are moving at velocity v, relative to the cat, the energy of the light emitted by the cat is calculated with the relativistic Doppler formula ($E = E_{flash}(1 + \frac{v^2}{2c^2})$, not derived for this proof). The total energy exchanged in our experiment is

$$E_{total} = \frac{1}{2}m_{beforeflash}v^2 - E_{flash}\left(1 + \frac{v^2}{2c^2}\right)$$

The same experiment could be repeated, with the cat emitting a flash of light before our ship departed when the energy associated with the flash of light is not affected by the relativistic Doppler effect. After we depart, we look at the cat and calculate its total energy as being:

$$E_{total} = \frac{1}{2}m_{afterflash}v^2 - E_{flash}$$

Because of the time-shift invariance of the universe (i.e., the same laws of physics apply regardless of the time that the experiment is conducted), the total energy of the experiment must not change regardless of when the cat emits its flash of light. Therefore,

$$\frac{1}{2}m_{beforeflash}v^2 - E_{flash}\left(1 + \frac{v^2}{2c^2}\right) = m_{afterflash}v^2 - E_{flash}$$

Rearranging, and noting that $m_{afterflash} - m_{beforeflash}$ is Δm, the mass equivalent of the light emitted by the flash of light:

$$\frac{1}{2}\Delta m v^2 = E_{flash} - E_{flash}(1 + \frac{v^2}{2c^2}) = E_{flash} - E_{flash} + E_{flash}\frac{v^2}{2c^2}$$

[28]Our approach is adapted from a YouTube video titled: "Einstein's Proof of $E = mc^2$", by minutephysics.

Dividing both sides of the equation by $\frac{1}{2}v^2$

$$\Delta m = \frac{E}{c^2}$$

Rearranging, we finally arrive at

$$E = \Delta mc^2$$

2.21 Proving the complete mass/energy equation

The relativistic equation for the energy of an object is,

$$E^2 = (mc^2)^2 + (pc)^2$$

Didn't we just prove, in the preceding section, Einstein's equation, namely $E = mc^2$? Why do we need these additions to the equation? The original equation does not show us the total contribution to the energy of the system coming from the kinetic energy of the object, moving at a speed v, from an observer's frame of reference. Basically, Einstein's equation, simplified for appeal to laymen, needs to be upgraded to meet relativistic specifications.

Any moving object gains mass. The kinetic energy of the object is the difference between the energy-equivalent of the moving object (i.e., the relativistically observed object) minus the energy-equivalent of the object at rest.

$$KE = mc^2 - m_0c^2 \qquad \text{Eq. 1}$$

Here, m_0 is the rest mass, and m is the relativistic mass (i.e., moving with respect to another reference frame). The moving object gains mass by a factor determined by the Lorentz transformation.

$$m = \frac{m_0}{\sqrt{1 - \frac{v^2}{c^2}}} \qquad \text{Eq. 2}$$

Therefore,

$$KE = mc^2 - m_0c^2 = \frac{m_0c^2}{\sqrt{1 - \frac{v^2}{c^2}}} - m_0c^2 \qquad \text{Eq. 3}$$

Let's try to find the relativistic energy of the mass that is moving at speed v with respect to the observer's frame of reference. Rearranging Eq. 2 and multiplying both sides of the equation by c^2,

$$mc^2\left(\sqrt{1 - \frac{v^2}{c^2}}\right) = m_0c^2$$

Squaring both sides of the equation,

$$(mc^2)^2\left(1-\frac{v^2}{c^2}\right)=(m_0c^2)^2$$

Evaluating the left side of the equation,

$$(mc^2)^2-\frac{m^2c^4v^2}{c^2}=(m_0c^2)^2$$

But mv is equivalent to the momentum, p, of the object. Substituting $p = mv$ and simplifying,

$$(mc^2)^2=(m_0c^2)^2+(pc)^2$$

Or, the observed relativistic energy-equivalent of the object (squared) is equal to the energy-equivalent of the object at rest (squared) $(m_0c^2)^2$ plus $(pc)^2$.

2.22 Proving kinetic energy of a moving mass is $\frac{1}{2}mv^2$

In high school physics, we learned the kinetic energy (KE) of a moving mass from first principles:

Kinetic Energy (KE) is force (F) acting over a distance, so,

$$KE = \int Fdx$$

Force is just mass, m, times acceleration, a.

$$F = ma$$

The acceleration at any moment of time can be expressed as the differential of velocity over a change of time, dv/dt. Putting it back into the equation for kinetic energy,

$$KE = \int m(\frac{dv}{dt})dx$$

But dx/dt is the velocity. Rearranging,

$$KE = \int mvdv$$

$$KE = Constant + \frac{1}{2}mv^2$$

We can infer that the constant must be zero (otherwise energy would not be conserved). This leaves us with the classic equation for kinetic energy,

$$KE = \frac{1}{2}mv^2$$

Does this formula apply when velocities are relativistic (i.e., near the speed of light)? Let's return to our relativistic equation for kinetic energy, (Eq. 2 in section titled "Proving the complete mass/energy equation"), and we'll see what we might discover.

$$m = \frac{m_0}{\sqrt{1 - \frac{v^2}{c^2}}}$$

So, the kinetic energy would be,

$$K.E. = 1/2m_0v^2 \cdot \frac{1}{\sqrt{1 - \frac{v^2}{c^2}}}$$

We can see that as the velocity approaches the speed of light, the denominator approaches zero, and the kinetic energy approaches infinity. Cosmic rays are the only mass-carrying particles known to travel at near-light speed. The rays consist mostly of free protons and can reach speeds of 0.9999999999999999999999 the speed of light. A single proton, traveling at this speed is thought to carry the kinetic energy of a baseball traveling at 90 kilometers an hour.[29]

[29]From an article published in *Universe Today* by Steve Nerlich, on December 24, 2015, titled "Astronomy Without A Telescope - OhMyGod Particles."

2.23 Proving the acceleration of waves is not infinite

"The career of a young theoretical physicist consists of treating the harmonic oscillator in ever-increasing levels of abstraction".
—Sidney Coleman

When you turn on your car's engine and step on the accelerator, it takes a while for the car to reach its cruising speed. Your car must pass through every speed in the interval between zero miles per hour and your final speed. If your cruising speed is 50 miles per hour, you can be certain that the car must have been traveling at 25 miles per hour at some prior moment.

When you switch on a flashlight, the light beam emitted will travel away from you at the customary speed of light. At the moment just before the switch is flipped, the beam's speed of light is zero. At the moment afterward, the beam's speed is approximately 186,000 miles per second. The beam of light emitted from the flashlight exhibits infinite acceleration. How does the light beam accelerate from zero to 186,000 miles per second in zero time?

It would seem that all waves achieve infinite acceleration and in the same manner. When you drop a stone in a pond, the waves close to the origin of the stone's splash will begin to move instantaneously, at a constant velocity (disregarding issues of friction and turbulence). The initial velocity of the wave and the final velocity of the wave are equivalent and are achieved instantaneously. The acceleration of the wave would appear to be infinite.

Does infinite acceleration actually exist? If force is mass times acceleration, then wouldn't an infinite acceleration require an infinite force? How is any of this possible?

Whenever our understanding of reality leads us to an impossible outcome, it's best to re-evaluate what we think we understand. In the case of wave movement, our intuitive understanding can be misleading. It is best to think of waves as simple harmonic oscillators that trace out waves in a disturbed field over time. The field may be air, water, or electromagnetic field, or any pervasive medium. The frequency of the wave is determined by the frequency of the oscillator.[30] The movement of the wave is a propagating disturbance in the field. Nothing is being accelerated.

For example, if a Tsunami wave originating near Japan were to travel across the Pacific Ocean, we can be certain that the wave that hits the California

[30]Harmonic oscillators (think of a spring bouncing up and down in place) are bounded systems (the spring stretches to a maximum, then snaps back, and repeats, and can be envisioned as the most stationary of stationary waves (i.e., a wave that oscillates up and down but doesn't go anywhere).

shore does not consist of Japanese water that was instantly accelerated by an infinite initial force, and magically carried to California. The wave that hits California is composed of California coastal water that was lifted by the propagating disturbance. The wave has propagated, but the water has simply oscillated (up and down).

We are taught (mistakenly) that when we flip a wall switch and our ceiling light suddenly flashes on, it does so because a completed electric circuit carried fast-moving electrons from the switch to the lightbulb. This is not the case. As it happens, free electrons in metal wires are sluggish, and move at a very slow rate (as slow as centimeters per hour). When a flipped switch completes an electric circuit, the change in voltage potential disturbs the electric field. The propagating disturbance of the field accounts for the nearly instantaneous circuit, and no electrons are actually accelerated and carried from the switch to the lightbulb.

In the case of waves, there is no wave acceleration and no mass acceleration; there is simply wave velocity.

2.24 Proving light carries momentum

As it happens, all systems containing mass also contain some electromagnetic waves (i.e., photons). These may come from some external source of light (such as the cosmic background radiation or a light bulb) or from light generated by the mass (e.g., from emissions associated with orbital transitions). Therefore, the total energy of an enclosed system must somehow account for the energy contributed by a stationary mass.

The total energy of any system must equal the energy of the objects at rest plus the energy produced by motion within the system.

$$E_{total} = E_{rest} + E_{motion}$$

If we have an object at rest, its contained energy is m_0c^2, where m_0 is its rest mass. But even as it sits still, we know that some energy is being emitted from the mass, adding to the total energy of the system.

We have shown that the complete energy equation for a system is

$$(mc^2)^2 = (m_0c^2)^2 + (pc)^2$$

For a system that is at rest, the motion of the system is accounted for by the motion of electromagnetic radiation. Therefore, the emitted electromagnetic radiation must equal $(pc)^2$, and this tells us that light, though it is massless, must have momentum!

We know that the energy of light is equal to $h\nu$. So,

$$(h\nu)^2 = (pc)^2$$

Taking the square root of each side of the equation and rearranging, we find the equation the momentum of light.

$$p = \frac{h\nu}{c}$$

Inasmuch as photons are massless, and the classical definition of momentum is $p = \text{mass} \times \text{velocity}$, we might guess that photons would have a zero momentum. Our guess would be wrong insofar as we can easily observe that photons do have momentum. For example, the photoelectric effect involves light (i.e., photons) knocking electrons from atoms, a reaction that obviously requires photonic momentum. We can also observe the momentum of light when we examine the tail of comets, which are always directed **away** from the sun, due partly to the effect of the momentum carried by the sun's electromagnetic radiation.

What gives light its momentum? We won't elaborate here, but it turns out that the momentum of fundamental particles comes from their spin, an intrinsic property of the wave nature of all types of particles. In the case of electromagnetic radiation, spin momentum is produced by the oscillating magnetic component of light (none comes from the oscillating electric component, which happens to cancel out to zero with each wave cycle).

2.25 Proving the impossibility of negative mass

"A serious and good philosophical work could be written consisting entirely of jokes".

—Ludwig Wittgenstein

Sometimes it seems as though everything in our universe has an opposite counterpart (e.g., positive and negative charges, positive and negative energy, positive and negative numbers, left and right-handed molecules, yin and yang, and so on). Is it possible that there is both positive and negative mass? Actually, no. Here's the proof.

1. Positive mass distorts space so that objects fall towards it (i.e., attraction). Conversely, negative mass distorts space so that objects move away (i.e., repulsion).
2. If a negative mass were placed near a positive mass, the positive mass would attract the negative mass and the negative mass would repel the positive mass, causing the positive mass to run away from the negative mass, with the negative mass chasing the positive mass, forever.
3. This would make possible perpetual motion machines (an impossibility).

   Therefore,
4. The universe must be composed of all positive mass or all negative mass.
5. But we have positive mass.

   Therefore,
6. There can be no negative masses.[31]

[31] The story is told of the college lecturer who claimed that that it is possible to form a positive assertion by combining two negative assertions. For example: "It is not true that he does not enjoy eating doughnuts." The lecturer further claimed that it is **impossible** to form a negative assertion by combining two positive assertions. A highly skeptical voice from the back of the classroom shouted "Yeah! . . . right!".

2.26 Proving both positive and negative energy must exist

"A mathematician is a person who thinks that when three people are supposed to be in a room but five came out, then you've got to send two people back into the room to empty it out".
—Origin unknown

Is it possible to have negative energy? Let's see.

1. We know that energy equals mass times the speed of light squared.
2. We know that mass is always positive (see section titled "Proving the impossibility of negative mass".) We also know that the speed of light squared is always positive.
3. The product of two positive numbers is positive.

   Therefore,
4. Energy must always be a positive quantity.

This cannot be right! We know that the existence of negative energy is necessary if we are to accept that energy is a conserved quantity in the universe.[32] Why is this so? If the net amount of energy in the universe is constant, then there must be a source of negative energy to counterbalance the positive energy. This being the case, where did we go wrong in our aforementioned proof that energy must always be positive?

Simply, our argument only addresses the energy contributed by mass and does not preclude the notion that negative energy may arise from sources other than mass.

We note that Einstein's full expression of the energy formula is:

$$E^2 = (mc^2)^2 + (pc)^2$$

In the **complete** energy equation, the value for energy can be either positive or negative, and its square (i.e., E^2) will be positive. The full expression of the relation between energy mass and momentum provides a way for energy to be negative. Just as well. If we accept that the net quantity of energy in the universe is zero (a requirement of the law of energy conservation), then we would need to accept that negative energy must exist, even if it has never been observed.

[32]See "Proving that the net amount of energy in the universe is zero".

2.27 Proving a phase change does not change wave equation observables

"Not everything that counts can be counted,
and not everything that can be counted counts."
—William Bruce Cameron

An easy way to think of wave phases, and to appreciate why the phase of a wave cannot be measured, is to imagine a clock face, with a minute hand but no other hands (i.e., no hour hand and no sweep-second hand). The clock face is graduated with 60 equally spaced ticks. The minute hand makes one full revolution every hour (i.e., the minute hand is a harmonic oscillator with a period of one hour). When we look at our clock, we see the minute hand at some location in the circle, but we have no way of knowing when its period began. It may have begun at tick 0 or tick 18 or tick 52. All we can say is that it is at its current location and that it will be at the same location in one hour. The clock **does** have a symmetry because it returns to its current location when rotated one hour. The location of the hand within its period is phase invariant because we cannot determine when the wave harmonic began. It has no absolute phase. Summarizing, the clock has a symmetry that is phase invariant.

If we wish, we can measure the difference in phase between our clock and any other clock. For example, if our clock is at tick 15, and another clock is at tick 30, then we know that the two clocks are 15 ticks out of phase. Equivalently, because the clock face is round, we could say that the two clocks are $\pi/4$ radians out of phase. Regardless, our measurement of the passage of time is wholly independent of either clock's phase.

Once we accept that the phase of a wave cannot be measured, we can infer that the phase cannot influence the outcome of an experiment.

1. Let's suppose the contrary assertion—that the phase of a wave **can** influence the outcome of an experiment.
2. If that were true, then we could design an experiment to determine the absolute phase of a wave (based on observing the outcome of the experiment that was influenced by the phase).
3. But we know that the absolute phase of a wave cannot be determined.
4. Therefore our original assertion is wrong.
5. Therefore the phase cannot influence the outcome an an experiment.

If the phase of a wave does not influence observables (e.g., the outcome of an experiment), then it obviously cannot increase the probability distribution for the wave equation (which is itself an observable outcome).

We can arrive at the same conclusion mathematically (if we care). A phase rotation of any wave equation, ψ, by an angle θ produces a new wave function, labeled as ψ', which has the following effect on its wave function and the wave function's complex conjugate:

$$\begin{aligned} \psi &\to e^{i\theta}\psi = \psi' \\ \psi^{\star} &\to e^{-i\theta}\psi^{\star} = (\psi')^{\star} \end{aligned} \tag{2.1}$$

The probability density of the wave function (i.e., the probability of finding a wave packet at any location) for a particular wave function, ψ is equal to $\psi^{\star} \cdot \psi$, where $\psi\star$ is the complex conjugate of the wave equation, obtained by changing the sign of the imaginary component of the equation. So, the probability density of our wave function before a phase rotation is

$$\psi = \psi^{\star}\psi$$

Under a phase rotation, the probability density of the new wave function, ψ' is

$$(\psi')^{\star}\psi' = e^{-i\theta}\psi^{\star}e^{i\theta}\psi = \psi^{\star}\psi$$

Because there is no difference in the probability density of a wave function before or after a shift in phase, we can conclude that phase shifts do not alter the observables of a wave function.

3

Universe and Cosmos

"Physics is the universe's operating system".
—Steven R. Garman

Cosmology is the science of the origin and development of the universe. Cosmologists describe the emergence of reality by applying the laws of physics to a set of initial conditions. Consequently, physicists need to have some level of confidence that the most fundamental operating principles of the universe are valid. In this chapter, proofs are provided for many of the foundational physical laws, including: a proof that the natural laws of the universe are conservative (e.g., the conservation of momentum and the conservation of charge), a short proof of the Heisenberg uncertainty principle, a proof that energy fluctuations must occur everywhere in space, a proof that there is no ether, a proof that entropy is not a conserved quantity, a proof of the Euler-Lagrange theorem, and a proof of Noether's theorem.

DOI: 10.1201/9781003516378-3

3.1 Proving the universe is conservative

Let's begin by assuming that the laws of physics are non-conservative.

1. If we were to perform any physical action under laws that were non-conservative, then we would expect some component of the universe to decrease as the result of the experiment.

   This is what is meant by a non-conservative law (i.e., something is not conserved as physical events occur over time).

2. But we assume that physical actions, and the laws under which they play out, are repeatable.
3. Therefore, each time a physical action occurs, we lose more and more of the non-conserved component.
4. Eventually, the non-conserved component would shrink to nothingness.
5. But the laws of physics apply to everything that happens.

   Therefore,

6. **Every** component of the universe would be exhausted as the laws of physics repeatedly come into play.

   Therefore,

7. The universe would shrink into nothingness if the laws of physics were non-conservative.
8. But we can observe that the components of the universe have not been shrinking.
9. Therefore, the laws of physics are conservative.

When we discuss processes that are non-conservative, we immediately think in terms of components of the system being lost or diminished (i.e., not conserved). We should also include processes in which some component is gained. In either case, the amount of the component at the end of the process is not equal to the amount at the beginning of the process (i.e., something is not conserved). To be complete, our proof should have included the case in which an experiment produces a net increase in a component. In this case, each repetition of the experiment produces more and more of the component until the universe is completely filled with the non-conserved component. The same logic applies. The universe we observe is conservative.

3.2 Proving conservation of momentum

The law of conservation of momentum derives directly from the homogeneity of spacetime.

1. Spacetime is homogeneous. That is to say that no position in spacetime is different from any other position in spacetime in the sense that the same physical laws apply everywhere, equally.[1]
2. Motion is a continuous translation through spacetime, and the same conditions of spacetime homogeneity apply.

   In the absence of applied external forces, as an object is translated from one position in space to another, the properties of the object and the forces acting upon the object do not change.
3. Therefore an object's motion (characterized by its mass and velocity, the two ingredients of momentum) does not change as it moves through space.

This is the basis for the law of conservation of momentum.

[1]Evidence supporting the homogeneity of spacetime is overwhelming. When we search space, we see a limited selection of celestial bodies (rocks of various sizes, stars, galaxies, galaxy clusters, black holes), all behaving themselves according to the same laws of motion, and all containing a similar assortment of atoms. The night sky looks about the same wherever and whenever we gaze. We are delighted when our powerful telescopes reveal some new wonder of the heavens, but every new wonder operates under the very same physical laws as our old wonders, regardless of its location.

3.3 Proving conservation of angular momentum

The proof of the conservation of angular momentum is much the same as our proof of the conservation of momentum. In addition to translation (i.e., shifting objects along a length from one point to another), our universe provides rotational symmetry about fixed points. As an object rotates, the physical status of the object at any angle of its rotation is conserved. In the absence of external forces, an object under continuous rotation will retain the same mass and the same angular momentum.

Let's take a moment to review some of the consequences of the conservation of translational momentum and angular momentum. Imagine that you are in a rocket ship moving at a constant velocity through empty space. For the sake of discussion, imagine that you have run out of fuel, and the rocket is simply coasting. When might you expect the rocket to come to a complete halt? The answer is "never". With nothing to impede the rocket's flight, it will continue along its current path, at its current speed, until the end of time. Forward momentum is conserved because no energy is being expended to counter its movement. Suppose you were to suddenly bump into a large, stationary object, with twice the mass as your rocket ship. Your rocket would stop in its tracks, and the larger object would suddenly fly off, much like a billiard ball hitting another billiard ball. At what speed would the more massive rocket fly? We know that momentum is conserved, and that momentum is mass times velocity. If your ship comes to a full stop after colliding with an object having twice the mass, then the more massive object will move away at half your former velocity, thus conserving momentum.

Now, imagine that you are spinning in space, at a constant angular velocity. Just as forward momentum continues for eternity, in empty space, so does the angular momentum of a spinning object. A sphere that is set spinning in space will continue to spin at a constant angular speed, forever. Suppose that the sphere suddenly, and mysteriously shrinks, without changing its mass, and without the exchange of energy. The angular momentum that was present in the larger sphere will be conserved in the smaller sphere.

Angular momentum is proportional to the mass of the rotating object multiplied by its radius. If the radius shrinks by half, the velocity of the rotating sphere will increase by a factor of 2, to conserve the angular moment that was present in the larger sphere. Likewise, if the radius shrinks by a factor of 100,000, then the rotational speed will increase by a factor of 100,000.

An increase in the speed of rotation will result when a spinning mass condenses in size. The observance of fast-rotating neutron stars tells us what to expect. In neutron star formation, a star that has a radius of a million kilometers will shrink down to a star with a radius of about 10 kilometers, a

100,000-fold difference. Angular momentum is conserved, and the neutron star spins with a velocity about 100,000 times the velocity of the parent star. Some neutron stars revolve hundreds of times per second.

FIGURE 3.1: When a large, spinning object is suddenly transformed into a small object having the same mass, its angular momentum, L, does not change (i.e., angular momentum is conserved). To preserve angular momentum, its angular velocity will increase by the same factor that the radius has decreased.

3.4 Proving energy is conserved

Time-shift invariance (also known as time translation invariance) is a symmetry of spacetime that asserts that the laws of physics do not change over time. An experiment performed on Monday will yield the same results on Tuesday, assuming that the experiment is conducted the same way at each time. Time-shift invariance may seem like an obvious trifle, of no special consequence, but it can be used to establish the law of conservation of energy, as follows:

1. If the laws of physics are time-shift invariant, then,
2. The properties of the universe that determine the outcomes of experiments must be conserved from one moment to another.
3. But all experiments involve energy (for motion, charge, forces acting over a distance).[2]

   Therefore,
4. The energy available from the universe must be time-shift invariant.

   Therefore,
5. The total energy of the universe must not change over time.

The last statement in the list is the law of conservation of energy.[3] At this point, we must caution that not everything that we have come to associate with the word "energy" is conserved. For example, force is not conserved. Though it is true that energy (in the form of work) is equal to the force applied over a distance, we should think of force as the result of applied energy (not as the source of energy). Better yet, we should probably think of force as the effect of a potential energy difference within a field. For example, the measured weight of an object is a force applied to a scale resulting from the potential energy of a gravity field. If we drop our object (sitting on our scale), it's weight instantly drops to zero. The weight (force) of the object is not conserved. The total energy of the system is conserved (as the sum of its potential and kinetic energies). Force is just a convenient measure of the observable effects of potential differences in fields. In earlier days, physicists

[2]We haven't strictly defined energy. For now, we can think of energy as whatever it takes to make things happen (e.g., to move something over a distance, to heat a gas from one temperature to another, to run an electrical motor, to keep the lights on.)

[3]In a later section, we will learn Noether's law, which asserts that wherever there is symmetry, there must be a conserved quantity associated with the symmetry. In the case of the symmetry of time-shift invariance, the conserved quantity is energy.

wedded to the concepts of physical forces opted to name fields for the forces they produced: i.e., electromagnetic force, strong force, weak force, and gravitation force. Would it be nitpicking to suggest that science would have been better served if physicists had named the fundamental fields by their energy potentials: electromagnetic potential, weak potential, strong potential, and gravitational potential?[4]

[4]In physics textbooks, you'll run across the term "conservative force". Don't be fooled into thinking that a conservative force is a type of force that is conserved. A conservative force is one that conserves the **energy** of the system under a closed movement, independently of the path taken (i.e., when an object ventures from point A to point B and back to point A with no net change in energy).

3.5 Proving mass is conserved

We can define mass as resistance against movement. Massless particles, such as photons, have no resistance to movement, and, in consequence, travel at the speed of light. Massive particles, such as protons and electrons, resist movement and travel at less than the speed of light. This definition conveys the idea that mass is not itself matter. Rather, mass is a property of matter. Therefore, a rock has mass, but a rock **is not a** mass. When we think of mass as a property that may have a relationship to other physical properties, we can see how mass may be a form of energy, as expressed in Einstein's mass-energy equivalence formula, $E = mc^2$. We prove Einstein's famous formula in another section. Using the formula, we can readily prove that mass is a conserved quantity.

1. Mass can be expressed in terms of energy, specifically $E = mc^2$

   Here, E is the total energy of the system we're studying, and m is the total mass of the system.

2. Rearranging the equation,

$$m = \frac{E}{c^2}$$

3. But we know that energy is a conserved quantity.

4. And we know that c is a constant.

   Therefore,

5. Mass, equalling E/c^2, must be conserved.

At this point, you might be wondering how mass or energy can be conserved if either can be transformed into the other. So, if we convert a mass, m, into energy (releasing mc^2 quantity of energy), then we have lost all of the mass and replaced it with energy, violating the laws of conservation of both mass and energy! To avoid this kind of dilemma, we ought to abandon the notion that mass is somehow converted into energy, or vice versa. We should accept that mass **is** energy. If we wished, we could do away with the concept of mass entirely, replacing it with the concept of E/c^2.

3.6 Proving charge is conserved

1. If the amount of charge in a charge-carrying particle were dependent upon an observer's frame of reference, then the amount of charge would vary as the charged particle and the observer moved relative to one another.[5]
2. If the amount of charge in a charge-carrying particle were to change, then the totality of charge in the universe would change, and charge would not be conserved.
3. But charge has energy, and a change in the total charge of the universe would produce a frame-dependent change in the total amount of energy in the universe.
4. But the total amount of energy in the universe is conserved (as per the law of conservation of energy).

   Therefore,
5. Particle charge is conserved and cannot depend on the observer's frame of reference (i.e., charge is absolute).

Just for fun, let's draw upon an advanced area of quantum physics to answer a simple but profound question. Would it be possible to suddenly create one unit charge somewhere in space, at the expense of a little energy? Charge conservation, on local and global scale, was formally proven by Hermann Weyl, using gauge theory, a mathematically sophisticated theory that we won't discuss here. From gauge theory, we learn that there are certain quantitative items in nature that cannot be directly measured. These are only understood and described in relative terms. One such item is voltage. It is impossible to measure the voltage (electrical potential) of a point in space. We can only measure the difference in voltage between points in space. Traditionally, the voltage at a point is measured by a comparison to the reference potential of Earth.

Given that voltage is only measured in relative terms, then let's consider a

[5]An inertial frame of reference is any reference frame that is not accelerating. This would include any reference frame moving at a constant velocity from zero up to (but not including) the speed of light in a vacuum. Any inertial reference frame can be matched with another reference frame moving at the same speed, so that each reference frame appears stationary to the other. In this book, whenever we mention "reference frames", we will be referring to initial frames of reference. The two working principles underlying all of special relativity are:
(a) The laws of physics are the same for all inertial frames of reference.
(b) The velocity of light is the same for all observers in inertial frames of reference.

Gedanken experiment in which we imagine that charge is not locally conserved. In this alternate universe, we can create a new unit of charge or destroy a unit of charge, at will. A unit of charge would be the charge carried by a single electron.

1. The energy carried by a unit charge is known to be equal to the charge times the voltage.

   ```
   Energy of charge creation = (charge produced) times
   (voltage at the point in space where the charge is produced)
   ```

   Hence,

2. If we measure the energy that was spent producing a single quantum of charge, then we would know that voltage at the point of charge creation.

   However,

3. Doing so would violate the gauge invariance of voltage, which is known to be a quantity that **cannot** be measured absolutely.

   That is to say that voltage is always a relative measurement of the differences of electrical potential at two points.

   Thus,

4. If the charge is not conserved, then gauge invariance is violated.
5. Gauge invariance cannot be violated.

   Therefore,

6. Charge is conserved, even for one lonely electron.

3.7 Proving the Heisenberg uncertainty principle

The Heisenberg uncertainty principle (named for Werner Heisenberg (1901-1976) arises from the wave nature of fundamental particles. Here is the mathematical expression describing the uncertainty principle:

$$\sigma_x \sigma_p \geq \frac{\hbar}{2} \,.$$

In the equation, x represents the position of a particle and p is the particle's momentum. The equation tells us that a change in the position of a particle times a change in its momentum can never be determined to a precision less than $\hbar/2$. Putting the equation aside for a moment, we can easily develop a logical argument indicating that the position and momentum of a particle can not be determined with exactitude.

1. Particles have wave properties.
2. A wave is not located at one position in space.
3. Measurements of a particle's position will differ depending upon where, in the wave's spread, a detector makes its measurement.

   That is to say that a particle's position must have a spread of detectable locations.
4. To detect a particle, the particle must interact in some way with the detector, and any such interaction would cause the particle to lose momentum.
5. Depending on how the wave interacts with the detector, the change of momentum may range from zero to the maximum momentum carried by the particle.
6. Multiple measurements of the momentum of the particle would yield a spread of values for the momentum

   Therefore,
7. We can never simultaneously determine the position and momentum of a particle with certainty.

Regarding proof of the uncertainty principle, we offer several. One proof is found in the section titled "Proving the canonical commutation relation". For the mathematically averse reader, we offer a watered-down "proof" in the section titled "Proving that there is a minimal particle frequency below which no energy is observable". In this section, we construct a proof using the Cauchy-Schwarz inequality.

Why do we have so many different proofs of the same principle? The Heisenberg uncertainty principle is one of several foundational concepts that help us understand the operating principles of reality. It has equal standing with the Pauli exclusion principle, the Schrödinger wave equation, and the theories of special and general relativity. Attaining new insight into the uncertainty principle is always a worthwhile endeavor.

The Heisenberg uncertainty principle applies to any two conjugate variables, (in this case, variables that are the Fourier conjugates of each other), so the uncertainty principle applies to either dual measurements for location and momentum or for energy and time. Let's consider position (x) and momentum (p). The Cauchy-Schwarz inequality for complex conjugate variables yields the following formula for the standard deviations(σ) of x and y.

$$\sigma_x^2 \sigma_p^2 = \langle f \mid f \rangle \cdot \langle g \mid g \rangle \geq |\langle f \mid g \rangle|^2 .$$

The complex conjugate of $\langle f \mid g \rangle$ is $\langle g \mid f \rangle$. We can show that,

$$|\langle f \mid g \rangle|^2 \geq \left(\frac{\langle f \mid g \rangle - \langle g \mid f \rangle}{2i} \right)^2 .$$

For complex conjugates,

$$\langle f \mid g \rangle = \int_{-\infty}^{\infty} f^*(x) \cdot g(x) \, dx,$$

And,

$$\langle g \mid f \rangle = \int_{-\infty}^{\infty} g^*(x) \cdot f(x) \, dx,$$

The asterisk in the equation denotes the complex conjugate. Now we insert the wave equations and evaluate.

$$\begin{aligned}\langle f \mid g \rangle - \langle g \mid f \rangle = &\int_{-\infty}^{\infty} \psi^*(x) \, x \cdot \left(-i\hbar \frac{d}{dx} \right) \psi(x) \, dx \\ &- \int_{-\infty}^{\infty} \psi^*(x) \left(-i\hbar \frac{d}{dx} \right) \cdot x \, \psi(x) \, dx\end{aligned}$$

This reduces to,

$$\begin{aligned}\langle f \mid g \rangle - \langle g \mid f \rangle &= i\hbar \cdot \int_{-\infty}^{\infty} \psi^*(x) \psi(x) \, dx \\ &= i\hbar \cdot \int_{-\infty}^{\infty} |\psi(x)|^2 \, dx \\ &= i\hbar\end{aligned}$$

Plugging this into the above inequalities, we get,

$$\sigma_x^2\sigma_p^2 \geq |\langle f \mid g\rangle|^2 \geq \left(\frac{\langle f \mid g\rangle - \langle g \mid f\rangle}{2i}\right)^2 = \left(\frac{i\hbar}{2i}\right)^2 = \frac{\hbar^2}{4}$$

Or, taking the square roots,

$$\sigma_x\sigma_p \geq \frac{\hbar}{2}.$$

This is the general Heisenberg uncertainty principle and can be applied to any pair of wave conjugates (including energy (E) and time (t).

From the Heisenberg uncertainty principle, we infer,

- If we know the position of a particle, then we cannot know its precise momentum.

  There must always be a degree of uncertainty expressed as $\sigma_x\sigma_p \geq \frac{\hbar}{2}$.
- Likewise, if we know the momentum of a particle, then we cannot know its precise location.

By framing the uncertainty principle into a limitation of physical measurement, it draws us away from the physical properties of waves, (whose features are spread out and not fully described by measured quantities), and, somewhat misleadingly, turns the uncertainty principle into a statement about the limitations of measurement. We should keep in mind, that the Heisenberg uncertainty principle does not only put a limit on the accuracy of physical measurement; it also puts a limit on what our measurements physically mean.

3.8 Proving the canonical commutation relation

In the process of deriving the Heisenberg uncertainty principle, we also derived another fundamental property of an operator for particular types of physical systems. The general equation for the canonical commutation operator is shown here.

$$\{\hat{x}, \hat{p_x}\} = \hat{x}(\hat{p_x}(\psi(x))) - \hat{p_x}(\hat{x}(\psi(x)))$$

The section titled "Proving the Heisenberg uncertainty principle" contains all that is needed to appreciate the mathematical steps shown below, and the discussion is not repeated here. Readers can skip ahead to the last few lines of this section to see how the derivation of the Heisenberg uncertainty principle yielded an important bonus: the value of the canonical commutator for Fourier conjugates (expressed as complex conjugates).

For complex conjugates:

$$\langle f \mid g\rangle = \int_{-\infty}^{\infty} f^*(x) \cdot g(x)\, dx,$$

and

$$\langle g \mid f\rangle = \int_{-\infty}^{\infty} g^*(x) \cdot f(x)\, dx,$$

The asterisk in the equation denotes the complex conjugate. At this point, we can insert the wave equations and evaluate.

$$\begin{aligned}\langle f \mid g\rangle - \langle g \mid f\rangle = &\int_{-\infty}^{\infty} \psi^*(x)\, x \cdot \left(-i\hbar\frac{d}{dx}\right) \psi(x)\, dx \\ &- \int_{-\infty}^{\infty} \psi^*(x) \left(-i\hbar\frac{d}{dx}\right) \cdot x\, \psi(x)\, dx\end{aligned}$$

This reduces to:

$$\begin{aligned}\langle f \mid g\rangle - \langle g \mid f\rangle &= i\hbar \cdot \int_{-\infty}^{\infty} \psi^*(x)\psi(x)\, dx \\ &= i\hbar \cdot \int_{-\infty}^{\infty} |\psi(x)|^2\, dx \\ &= i\hbar\end{aligned}$$

The value *ih* is a fixed value that describes a particular relationships between any two physical physical properties that happen to be the Fourier transforms of one another (i.e., Fourier conjugates). This relationship, as shown here for locations (x) and momentum (p) is known as the canonical commutator and shown:

$$\{\hat{x}, \hat{p}_x\} = i\hbar$$

Incidentally, the (absolute) average of the possible outcomes of position and location, gives us $\hbar/2$, their Heisenberg uncertainty.

3.9 Proving energy fluctuations must exist everywhere in space

The Heisenberg Uncertainty Principle tells us something about the nature of observations (fuzzy, at best), but does it tell us anything about the "contents" of "empty" space? Let's look again at the uncertainty associated with energy and time.

$$\triangle E \cdot \triangle t \geq \frac{\hbar}{2}$$

It would seem that we can never be certain that the energy, at any location, is zero. If we had a certain level of zero, then the product of the uncertainty in energy and time would also be zero, but the product of the uncertainty of energy and time must be at least equal to $\frac{h}{4\pi}$. Therefore, we can never be certain that any location in our universe, at any time, has a value of zero!

If this were the case, then we must see random fluctuations in the energy level, in every volume of space, to account for our uncertainty. Put another way, non-zero energy must arise from the vacuum of space, and the length of time during which this energy exists must not be zero; otherwise, the Uncertainty principle would be violated.

Furthermore, because there is a direct relationship between energy and mass, and because fluctuations of energy must occur, we can infer that particles (i.e., field quanta) will arise spontaneously from the vacuum of space and these particles must exist for a short but finite length of time (such that $\triangle E \triangle t \geq \frac{\hbar}{2}$). These particles are too small and too short-lived to be observed and are called "virtual" particles. Every field has its own non-zero fluctuations and produces field-specific virtual particles (e.g., virtual quarks, virtual photons).

Finally, we can speculate that in a seemingly infinite and eternal universe, we might expect to encounter random energy fluctuations that are extremely large. Theorists speculate that singularities may arise when a large spontaneous energy fluctuation occurs within a very small volume.

Before we are completely carried away with the cosmic consequences of the humble Heisenberg uncertainty, let's touch base with terra firma and review the experimental evidence that provides some credibility to our wild hypotheses. As it happens, there are two phenomena that confirm that energy arises from a vacuum. These are the Casimir effect and the Lamb shift.

In 1948, Hendric Casimir predicted that a vacuum, devoid of matter and energy and shielded from external energy by two metal plates, will produce energy fluctuations of its own. In 1997, Steve Lamoreaux conducted the experiment and he showed that the vacuum between the plates created a net force (attractive or repulsive) that measured within 5% of Casimir's theoretical calculations. **Thus, something can arise from nothing.**

The Lamb shift is a quantum phenomenon of atoms that was named after Willis Lamb who, along with Robert Retherford, conducted an experiment that demonstrated vacuum fluctuations occurring within atoms. According to predictions based on the Dirac equation, the 2s1/2 and the 2p1/2 (the spin 1/2 electron in the s level of the second orbital shell and the spin 1/2 electron in the p level of the second orbital shell) energy levels of the Hydrogen atom should be equal. They are not equal, and it turns out that vacuum fluctuations (i.e., the Casimir effect) nicely account for the differences in their energies.[6]

Trust in the Heisenberg uncertainty principle is based upon our confidence of the wave nature of matter. If matter cannot be described as waves, then the underpinnings of the Heisenberg uncertainty principle (ie., the Schrödinger wave equation and our understanding that position and momentum are complex conjugates of one another under Fourier transformation) would vanish, and so too would the Heisenberg uncertainty principle. Einstein held the opinion that every particle has an exact position and momentum at any moment in time. If Einstein was correct, then the Heisenberg uncertainty principle tells us more about the limitations of wave functions, than limitations of spacetime.

[6]For the original Lamb and Retherford manuscript, see Lamb WE, Retherford RC. Fine Structure of the Hydrogen Atom by a Microwave Method. Physical Review 72:241-243, 1947.

3.10 Proving a particle's location is measurable only to 1/2 its Compton wavelength

"The solution to a problem changes the nature of the problem".
—John Peers

Imagine, for a moment, that you are a particle such as an electron, a proton, or a neutron. Now, ask yourself, "If I were a light wave, rather than the particle that I am, then what would be my wavelength? The answer to this wildly hypothetical proposition is "The Compton wavelength"! [7]

Every particle has a specific resting mass, and this resting mass is associated with a specific energy. Likewise, every electromagnetic wave, though it has no mass, is associated with a specific mass-equivalent energy. The Compton wavelength is the wavelength of light whose energy is the rest energy of the particle. Expressing these relationships in mathematical terms, we see that a particle of mass m has a rest energy of $E = mc^2$, Einstein's famous equation. Furthermore, the energy of an electromagnetic wave is $h\nu$ (i.e., its frequency, ν, times the Planck constant, h.

When we look for an electromagnetic frequency that has an energy equal to the energy of a particle with mass m, we have:

$$E = h\nu = mc^2$$

Therefore, when the energy of an electromagnetic wave and a particle are equal, then the corresponding frequency of the wave must equal mc^2/h. Furthermore, since the frequency of an electromagnetic wave is related to its wavelength by the equation $\nu = c/\lambda$, then the corresponding wavelength (i.e., the Compton wavelength) is h/mc.[8] By dividing both sides of the Compton wavelength equation by 2π we have the reduced form of the Compton wavelength, expressed using the reduced form of Planck's constant, as $\hbar/mc$. The purpose of doing so permits us to express the Compton wavelength in terms of radians, rather than 2π radians.

What's the point of all this? As it happens, we can learn a great deal from the Compton wavelength of different particles. With it, we can determine the particle's radius, and we can learn how photons were scattered in the early

[7] Named for Arthur Compton (1892-1962).

[8] When the author was barely surviving undergraduate physics, a popular joke at the time involved greeting a classmate with the question, "What's new?", and the classmate would reliably answer E/h. Not so funny, in retrospect, but it serves as a sturdy mnemonic for the relation between frequency, ν and the energy of electromagnetic radiation.

years following the Big Bang, when the universe was confined to a dense hot plasma. Just for fun, let's show that we can never determine the position of a particle to within half of the particle's Compton wavelength.

To refresh, the Heisenberg uncertainty principle asserts that we can never measure a change of momentum $\triangle p$ times a change of position $\triangle x$ to within $\hbar/2$, where $\hbar$ is the reduced Planck constant.

Expressed as an equation, we have:

$$\triangle x \, \triangle p \geq \frac{\hbar}{2},$$

Or,

$$\triangle x \geq \frac{\hbar}{2 \, \triangle p},$$

Let's imagine that the momentum of the particle is as great as it can possibly be, and maybe a tad more. This would mean that the particle moved at the speed of light, giving it an upper limit of momentum to be mc. Let's put this upper limit into the equation to determine what this does to our limit on finding the position of the particle with certainty.

$$\triangle x \geq \frac{\hbar}{2mc} = \frac{1}{2} \times \frac{\hbar}{mc}$$

But we have previously shown that $\hbar/mc$ is just the reduced Compton wavelength.

So, the position of a particle can only be measured to a precision of one-half of the reduced Compton wavelength of the particle. This conclusion helps us to understand some of the properties of dark matter, hypothesized to account for the clustering of galaxies and the increase in the rotational speed of galaxies over time.[9] For example, the predicted mass of a hypothetical dark matter particle is $10^{-22}eV$, a number that really does not mean very much to us. However, a dark matter particle with this mass will have a Compton wavelength of about 1.3 lightyears, meaning that we could never know the position of a dark matter particle to anything under 0.65 light years. Viewed in terms of its Compton wavelength, we know that a particle of dark matter has no discernible position, no structure, and a limited ability to interact

[9]Not all physicists are confident of the existence of dark matter. A theory known as Modified Newtonian Dynamics, first developed by Mordehai Milgram (1946-), predicts galaxy condensation without hypothesizing the existence of dark matter. See Milgram M. A modification of the Newtonian dynamics as a possible alternative to the hidden mass hypothesis. Astrophysical Journal 270:365-370, 1983.

with just about anything. It comes as no surprise that we have so far failed to detect dark matter particles.[10]

[10]Incidentally, the term "dark matter" is a misnomer. Dark matter cannot block out light waves and is not dark. A better name would be "transparent matter".

3.11 Proving black holes exist

In first-year physics, students are asked to compute the speed with which an object must travel for it to escape the Earth's pull of gravity. We could modify the question as follows "What mass must Earth have to ensure that a rock of a given speed will fail to escape Earth's gravitational pull?" If we substitute light for the rock, and knowing that the speed of light is a constant, we can rephrase the question as, "What mass do we need to ensure that an object moving at the speed of light cannot escape the gravitational pull of the mass? The answer to this question provides us with the minimal requirements for a black hole, insofar as a black hole is defined as an object from which nothing can escape.

The heuristic equation for escape velocity, which does not take into account relativistic effects, is

$$v_{escape} = \sqrt{2GM/R}$$

Here, v is the escape velocity, G is the gravitational constant, M is the mass of the object exerting a gravitational pull and R is the distance from the center of the mass to the escaping object.

What would happen if $\sqrt{2GM/R}$ (the right side of the equation) was greater than the speed of light? In this case, no matter could possibly escape from the mass because matter cannot achieve a speed greater than the speed of light. In addition, no light could escape from the mass for the same reason (i.e., light cannot travel faster than the speed of light). We use the term "black hole" to describe a mass from which nothing can escape within a radius R. We use the term "event horizon" to describe R, the distance between the center of the black hole and the object that is attempting to escape. Whatever light and mass are present within the event horizon would not have the speed needed to escape the gravitational attraction of the black hole.

We can see that the larger the mass of the black hole, the greater the radius of the event horizon. A black holes with extremely small mass would have a very small event horizon. There is no upper limit to a black hole's mass. In theory, a single black hole may engulf all of the matter in the universe. There is also no lower limit to the size of a black hole. Any mass, when confined to a small enough volume, will form a black hole. We might add that as the black hole approaches the zero limit in mass, so will the radius of its event horizon. In this limiting case, the black hole would be unobservable.

We can take a moment to discuss how scientists have proven that black holes, objects that cannot be seen, actually exist.[11] Observations from which we can infer the existence of black holes generally fall into one of two categories:

[11] Because black holes, by definition, do not emit light, they cannot be directly observed.

1. Observances of strong gravitational effects in a spatial location that does not contain visible objects.

   The gravitational effect of black holes is always large insofar that the gravity of black holes must be sufficient to retain light. Rapidly moving clusters of stars orbiting around "nothing" suggests a black hole at work.

2. Observances of the release of great amounts of energy surrounding an area containing no visible object.

   As matter comes near a black hole, it is accelerated by gravity, and this acceleration increases the kinetic energy of particles, producing tremendous heat in the form of high-energy radiation. A glowing disc of hot gas around an empty space suggests a black hole. Jets of high-velocity matter and fast-moving magnetic fields, emanating from an apparently black spot in space, are consistent with the presence of a large black hole that is spinning.

3.12 Proving the universe is expanding

"Man lies 'suspended between two infinities.'"
—Blaise Pascal, referring to the infinitely small and the infinitely large

On occasion, a simple-minded argument-by-absence can be helpful. For example, cosmologists have, through the past several centuries, wondered whether the universe is expanding or contracting. Various individuals, including Mark Twain and Emmanual Kant, have been credited with inventing the following argument demonstrating that the universe must, in fact, be expanding.

1. If the universe were of a fixed size, or if the universe were shrinking, then all of the light being constantly generated by the countless stars in the universe would be retained within its limited volume.
2. Over time, the night sky would become full of light; blinding us with its power.
3. But the night sky is not getting brighter over the ages.

   All we see in the moonless night sky is darkness punctuated by scattered points of feeble light.

   Therefore,
4. The universe is expanding.

   We shall probably never observe a white night.

Strictly speaking, the observation of night-time darkness does not rise to the level of proof. Nonetheless, the dark night sky is consistent with the hypothesis that the universe is expanding, and is inconsistent with the argument that the universe is contracting. A stronger argument for the expansion of our universe came early in the twentieth century, when astronomers found that wherever we search, galaxies are receding from us. This observation, discovered by Vesto M. Slipher in 1912, was confirmed, in 1929, by Edwin Hubble, further showed that the speed at which galaxies were moving away from Earth was directly related to their distances from Earth (i.e., the further the galaxy, the greater its speed).

3.13 Proving the Big Bang

"There is a crack,
a crack in everything.
That's how the light gets in".
—Leonard Cohen, from the song *Anthem,* in his album, *The Future*

About 13.8 billion years ago, there was a cosmological event that has been likened to the starting pistol that set the observable universe in motion. We base our understanding of the Big Bang on cosmological observations of distant space, which provides a backward look at our universe as it developed over time. Arguments supporting the Big Bang theory of the universe include the following:

- As previously noted (see section titled "Proving that the universe is expanding"), wherever we look in the night sky, we see galaxies receding from us, with the most distant galaxies moving at the greatest speed. Such observations led to Alexander Friedmann's suggestion, in 1922, that the universe is expanding. Georges Lemaître, in 1927, proposed that the expansion of our universe must have begun at a moment in the distant past when all matter was confined to a single point; then suddenly released.[12]
- Hydrogen and Helium account for nearly all of the atoms in the universe. This suggests that there was a moment when matter existed as a condensed hot plasma of protons, neutrons, and electrons, as would be required for the Big Bang theory.
- There is a cosmic background of radiation that is present, in all directions, and this background looks about the same everywhere. This suggests that the aforementioned hot plasma cooled to the point where electrons were captured by light nuclei (protons or proton-neutron combined), releasing confined photons, in one burst. The released photons could then fill the universe with radiation. Because the current cosmic background radiation is the same wherever we look, we can assume that the photon release event filled the condensed early universe, and has continued to fill the universe, as the universe has expanded to its current size.[13]
- Space telescopes are basically machines that view the past, permitting us to view the development of the observable universe, beginning with the Big Bang. When astronomers announce that they have detected a star that is located 5 billion light-years from Earth, we are really saying that we have

[12]Decades later, in 1981, Alan Guth proposed inflation wherein from 10^{-36} seconds to 10^{-32} seconds following the Big Bang event, our universe expanded 10^{78} times.

[13]For further details of the cosmic background radiation, see section titled "Proving most of the observable energy of the universe is the cosmic microwave background"

received light that has traveled 5 billion years after being emitted by a star. What the astronomers see is what the star looked like 5 billion years ago. The current location of the star, and whether the star even exists at the current moment, is unknown to us. By viewing stars at increasing distances from Earth, we can reconstruct a sequence of events moving backward in time. Consequently, astronomers have a fairly good understanding of how the universe changed over time, leading back to a moment when the universe was hot and dense and tiny.[14]

- Despite our best efforts, we cannot find stars more distant from Earth than about 13.5 billion light years. The light from a star coming into existence 13.5 years ago would require 13.5 billion years to reach Earth. The absence of stars at a distance greater than 13.5 billion light years would suggest that no stars existed more than 13.5 years ago. Allowing some time for the first stars to form following a Big Bang event, would suggest that the Big Bang probably occurred about 13.8 billion years in the past. Conversely, if we found a star that formed, say, 20 billion years ago, then we would have proof that the Big Bang, as currently conceived, must not have occurred.[15]

We should note that our arguments for the Big Bang are based upon our current interpretations of numerous observations. As always, our interpretations may be incorrect (i.e. we may misunderstand what we think we've seen), and our observations may be premature or incomplete (i.e., additional observations may change the validity or meaning of the earlier observations). The Big Bang, in particular, is a particularly troublesome theory inasmuch as it creates more questions than it answers, including,

- Why was there a Big Bang?
- What, if anything, preceded the Big Bang?
- How did the Big Bang produce what we observe as the reality of our universe?
- Were there other Big Bangs that we should know about?
- Might we expect the next Big Bang to occur anytime soon?

---

[14]Soren Kierkegaard, commenting on one of the ironies of reality, wrote that "life can only be understood backward; but it must be lived forwards."

[15]Because the universe itself is expanding, the distance between Earth and the stars we observe is much greater than the distance traveled by light emanating from the stars. When we say that no stars have been found more distant than about 13.5 billion light-years from Earth, we are stipulating that the star's light received on Earth departed from the star no earlier than 13.5 billion years ago.

3.14 Proving the net average amount of energy in the universe is zero

Let's use the principle of energy conservation to prove that the net amount of energy in the universe is zero and that negative energy must exist.

1. Energy is conserved over time.
2. All of the observable energy in the universe today was present at the moment of the Big Bang.

   In fact, most of the electromagnetic energy observed in the universe today is found in the cosmic background radiation released sometime after the Big Bang, and most of the atoms in the universe today are the same atoms that were produced when the cosmic background radiation was released.
3. Under the assumption that nothing existed before the Big Bang, we can infer that the net energy prior to the Big Bang was zero.
4. Energy conservation requires that there was the same amount of energy in the universe before the Big Bang as there was during the Big Bang and after the Big Bang.

   Therefore,
5. The net amount of energy in the universe today is zero.

   Therefore,
6. There must be negative energy in the universe to balance the observable positive energy.

Though negative energy may exist, we have proven that it is impossible to have negative mass (see section titled "Proving that negative mass is impossible"). This being the case, and knowing that the energy equivalent of a rest mass is $E = mc^2$, we can infer that energy derived from mass is always positive energy. Therefore, we must look to sources other than mass to find negative energy. As it happens, much has been written on the subject. Antimatter (as per Paul Dirac), Black holes (as per Stephen Hawking), virtual particles, and gravitational potential energy, are all candidate sources of negative energy. Science fiction buffs will tell you that worm holes and warp drives owe their existence to negative energy. Are such speculations valid? We should not judge them too negatively.

3.15 Proving there is no ether

"This isn't right. This isn't even wrong!"
—Wolfgang Pauli (when criticizing a colleague's half-baked theory)

Nineteenth-century scientists assumed that light was carried by an invisible substance known as the ether.[16] They reasoned that if forces such as light and gravity acted upon objects separated by a distance, then there must be some physical medium that carries those forces through the intervening space. This pervasive medium was called "the ether."

To prove experimentally that their assumption was true, two physicists working near the end of the 19th century conducted a clever experiment designed to prove the existence of the ether, conclusively. Their methodology was based on the following line of thinking:

1. If there is an ether, then the Earth must move through it as Earth circles the sun.
2. As the Earth moves through the ether, it must produce a sort of "drag" or "ether wind" diminishing the speed of light that happens to travel against the direction of the wind.
3. Two light beams, originating at the same moment, but traveling at right angles to one another, will experience different levels of ether wind.
4. Hence, the two beams will travel at different speeds.
5. The difference in speeds of the two beams will become apparent in a precision interferometer, even when the differences in speed are quite small.
6. The finding that light travels at different speeds, depending on the direction of drag, produced by the Earth's movement through the ether, would be iron-clad proof that the ether exists.

Famously, Albert Michelson (1852-1931) and Edward Morley (1838-1923) found **no difference** in the speed of the two light beams, regardless of their respective paths through the ether.[17] Their experiment, published in 1886,

[16] Do not confuse the ether of space with the volatile liquid anesthetic of the same name.

[17] Michelson AA, Morley EW. On the relative motion of the Earth and the luminiferous ether. Am J Science 34:333-345, 1887.

challenged the prevailing faith in the ether, and opened the door for Einstein to introduce the world to relativity, 18 years later.[18]

Aside from the Michelson-Morley experiment, we have good reason to suspect that the ether envisioned by 19th-century scientists could not possibly exist.

1. The universe is continuously expanding

   For a discussion of this topic, see section titled "Proving the universe is expanding".

2. As the universe expands, the space between objects increases.

   The fundamental particles that occupy the universe do not change size or increase in number during the expansion of space.

3. Space can expand indefinitely, without violating conservation laws, because space is an abstraction that indicates the time required to travel between points.

4. If an ether exists, it must increase to fill the geometry of expanding space.

   For example, if space increases its volume 2-fold, then twice as much ether must be created to fill the added volume. The ether cannot simply attenuate as space expands. Doing so would alter its function as a carrier of electromagnetic radiation and other physical fields. This would result in an inhomogeneous spacetime where the laws of physics change as the universe enlarges.

5. Ether, unlike space, was believed to be a physical presence. As such, ether would require an energy source for its creation.

6. The unlimited expansion of the universe would require an unlimited source of energy to produce an unlimited quantity of ether.

7. Energy is a strictly conserved quantity, and cannot account for a process that keeps everything in the universe unchanged **except** for replenishing an unlimited supply of ether.

---

[18]Michelson AA, Morley EW. Influences of motion of the medium on the velocity of light. Am J Sci 31:377-386, 1886.

Therefore,

8. The ether cannot exist.

In the section titled "Proving that nothingness has physical properties," we shall prove that the universe does have an ether; it's just not the ether that 19th-century scientists had in mind.

3.16 Proving entropy is not a conserved quantity

Entropy can be discussed under any of the following constructs:

- As the measure of a system's disorder.
- As the measure of the combinations of ways that the contents of a system can be rearranged.

  A crystal has relatively little entropy, compared to a gas.
- As the amount of work that can be extracted from a system (by moving heat energy around).
- As the numeric embodiment of the second law of thermodynamics.

  Thermal reactions that are irreversible always increase the system's entropy.
- As the information needed to describe a system (including the quantum states of every atom in a system)

  The system's information, as formulated below, takes all of the states of the system, x_i, determines the probability of each state, $P(x_i)$ and expresses the summation for all n possible states:

  $$\mathrm{H(X)} = -\sum_{i=1}^{n} \mathrm{P}(x_i) \log_b \mathrm{P}(x_i) \qquad \text{Information Entropy formula}$$

- As a measure of the possible outcomes of an event

  An event that has only one or several outcomes is a low entropy event. An event that has many possible outcomes is a high entropy event.
- As the cause for the forward direction of time

  Insofar as both entropy and time both seem to be continuously increasing, an increase in either one would seem to tell us that the other quantity is also increase. However, it's a logical stretch to infer that entropy **causes** time to move forward.
- As the measure of a thermodynamic state, determined by the following formula, wherein S is entropy, Q is heat, and T is temperature

  $$dS = \frac{\delta Q}{T}.$$

A simple and practical way of thinking of entropy is as a number that we can calculate that tells us something about how much we would need to know if we wanted to fully describe a system. Increasing the number of components of a system would increase its entropy, as would increasing the degrees of freedom by which those components may move, as would increasing the space in which we might find those components. Compressing the universe to a single point in space, containing no distinguishable objects, would decrease the entropy of the universe. Processes that are reversible are not associated with a change in entropy insofar as the state of the system must be equivalent before and after the process is completed (that is what it means to be reversible). In a static universe, wherein every process is reversible and the size of the universe is fixed, the number associated with the entropy of the universe would be constant over time (i.e., conserved).

We get a practical taste of entropy by playing with the calculated entropies of image files. A near-empty image file has a calculated entropy very close to zero.[19]

An image composed of pixels each having a randomly selected gray level between 0 and 255 will require a great deal of information to fully describe the full image. It's calculated entropy will be very close to the maximal, 1 (actually, 0.999893).

When we combine two distinguishable images by overlaying one atop the other, we always see an increase in the calculated entropy (i.e., the entropy of the combined image is always greater than that of either of the individual images). Overlaying one image atop another is an irreversible process (unless we add information to the system).

However, when we combine an image with itself the entropy of the combined image is exactly equal to the entropy of each identical image alone. Overlaying an image on itself is a reversible process. Combining an image with itself is analogous to the mixing experiment wherein two volumes of identical atoms are combined.[20] In both cases, there is no change in the entropy of the system.

In summary,

1. When we look at the entropy associated with physical processes, we see that there are processes that do not change the entropy of

---

[19]The open source image manipulation program, ImageMagick, provides a command-line utility that yields the entropy of an image. Once ImageMagick is installed, a typical c prompt command and return would appear as follows:
identify -format "entropy %[entropy] filename %f" black.jpg
entropy 0.0471023 filename black.jpg

[20]The mixing experiment was discussed in section titled "Proving the identicality of fundamental particles of a kind". When gases consisting of identical molecules are mixed together, there is no change in entropy (unlike the situation when different gases are mixed together). From this observation, we infer that all of the individual atoms of any pure gas are identical to one another.

the system, and there are processes that increase the entropy of the system.

2. We never observe physical processes that result in the reduction of the entropy of the system.

   That is to say, systems do not spontaneously organize themselves to lower entropy states.

   Therefore,

3. We infer that entropy never decreases.
4. If entropy never decreases,

   Then,

5. Entropy is not conserved.

FIGURE 3.2: A simple image (black background with a small triangle of white) requires very little information to describe the image. Consequently, it has a calculated entropy near zero (actually, 0.0471023).

FIGURE 3.3: An image of randomly selected grey-value pixels. It's calculated entropy is nearly 1.0 (more precisely, 0.999893).

FIGURE 3.4: Two images superimposed. The entropy of the combined image is greater than that of either of the component images alone.

3.17 Proving the Euler-Lagrange equation

Here is the Euler-Lagrange equation:

$$\frac{\mathrm{d}}{\mathrm{d}t}\left(\frac{\partial L}{\partial \dot{x}_0}\right) = \frac{\partial L}{\partial x_0}$$

On first inspection, this equation doesn't seem to bear any profound significance, but physicists rely upon this equation whenever they model the laws and the symmetries that represent physical reality. For us to appreciate the relevance and power of the Euler-Lagrange equation, we first need to introduce the Lagrangian, and its so-called action integral. The Lagrangian is a mathematical function that encapsulates the forces operating within a space. It is suited for solving sets of equations where the forces are conserved.

In mathematical terms, the Lagrangian is the difference between the kinetic energy of the system (i.e., the sum of the kinetic energy of all the parts of a system) and the potential energy of the system. In effect, the Lagrangian describes the physical system.[21]

$L = T - V$ The Lagrangian, T=kinetic energy, V=potential energy

When we speak of symmetries under Noether's theorem, we are referring to symmetries of the Lagrangian (see section titled "Proving Noether's theorem"). That is to say that we are dealing with those symmetries that leave the Lagrangian unchanged. The "Action" of the Lagrangian is the Lagrangian's integral over time.

$$S = \int_{t_1}^{t_2} L(x, \dot{x}, t)\,\mathrm{d}t \qquad \text{S = Action}^{22}$$

[21]We are taught that potential energy is the "stored" energy of a system. This definition is highly misleading. In technical terms, potential energy is the difference between the energy of an object in its location and its energy at a reference location. In practical terms, potential energy is the energy that would be spent to return the object to its original state after some kinetic event. In the case of a rock dropped from the top of a cliff, the potential energy is the energy required to lift the rock from the cliff bottom to the cliff top. It might be of value to replace the term "potential energy" with "restorative energy", the energy needed to restore the system to its original state. In a fully reversible process, the restorative energy equals the kinetic energy.

[22]Note that S is a functional, not a function. Its value is determined by the values of all the functions in the integrand, over all the points of time in the evaluated interval.

The action has an extreme value (i.e., a maximum, a minimum, or a saddle-point) at locations where an infinitesimal variation in the action is zero. [23] As shown here,

$$\delta S = \delta \int_{t_1}^{t_2} L(x, \dot{x}, t) dt = 0 \qquad \text{Least Action}$$

Why is this the case? When a function (or a functional, in the case of the Action) is at a minimum or maximum, or saddlepoint, it is at a stable position, where there is no accelerated movement, and this calmness extends to the values in its immediate vicinity. So an infinitesimal variation in the Action at its extremum will not change the Action value (i.e., its derivative is zero).

To describe the laws of motion in a system having conserved forces, we must find the function $x_0(t)$ that produces Least Action (i.e., a function whose action will be an extremum, and whose derivative is zero. To review, at the Action's extremum, the Action will not change over small variations in time, so that its derivative, $\delta S = \delta \int_{t_1}^{t_2} L(x, \dot{x}, t) dt$ will equal zero. Where the Action does not change over time its Least Action becomes a conserved quantity.[24] We can also point out that when we discuss physical systems, actions involve motions (changes of position, the "x" variable, over time (the "t" variable). So we can intuit here that discovering $x_0(t)$, the function that produces the Least Action, will lead us to the equations of motion in a conserved system.

Physicists find $x_0(t)$ by using the Euler-Lagrange equation.

$$\frac{\mathrm{d}}{\mathrm{d}t}\left(\frac{\partial L}{\partial \dot{x}_0}\right) = \frac{\partial L}{\partial x_0} \qquad \text{Euler-Lagrange equation}$$

Before we get ahead of ourselves, we must prove that the Euler-Lagrange equation is true. Our task seems daunting, but all we need for our proof is the chain rule (see section titled "Proving the chain rule") and integration by parts (see section titled "Proving integration by parts").[25]

Let's begin by assuming that $x_0(t)$ is a function that produces a Least Action for S (i.e., wherein $\delta S = 0$. Now consider some other function $x_a(t)$ such that:

$$x_a(t) = x_0(t) + a\beta(t)$$

[23] To appreciate the concept of a saddlepoint, imagine that we have a saddle on a horse, and we are placing a small pebble on the saddle. There may be points of stability on the surface of the saddle from which small nudges in position will not cause a pebble to fall down or rise up. These would be the saddlepoints.

[24] Remember, a conserved quantity is anything that does not change over time.

[25] Here, we loosely follow the proof described by David Morin in "Introduction to Classical Mechanics With Problems and Solutions", Cambridge University Press, Cambridge, UK, 2008.

Here, a is a number and $\beta(t)$ is a function whose endpoints at t_1 and t_2 yield $\beta(t_1) = \beta(t_2) = 0$.

Now, let's return to our definition of the Action:

$$S = \int_{t_1}^{t_2} L(x, \dot{x}, t)\, dt \qquad \text{S = Action}$$

We recall that the Action is a functional (not a function) and depends upon all of the possible functions of $x_{anything}(t)$ that satisfy the Lagrangian equality. So we can take the partial derivative of the Action, for the changing functions, using a variable, a, in $x_a(t)$ like so:

$$\frac{\partial}{\partial a} S[x_a t] = \frac{\partial}{\partial a} \int_{t_1}^{t_2} L(x, \dot{x}, t) dt = \int_{t_1}^{t_2} \frac{\partial L(x, \dot{x}, t)}{\partial a} dt$$

But L is a function with three variables $L(x, \dot{x}, t)$, so in our equation, we have:

$$\partial L = \partial L(x_a) + \partial L(\dot{x_a}) + \partial L(t)$$

But we are looking for some $x_a t$ that conserves the Lagrangian over time, so we can drop the last term of the right side of the equation to give us,

$$\partial L = \partial L(x_a) + \partial L(\dot{x_a})$$

Plugging this into our equation,

$$\frac{\partial L(x, \dot{x}, t)}{\partial a} = \frac{\partial L(x_a)}{\partial a} + \frac{\partial L(\dot{x_a})}{\partial a}$$

The chain rule tells us that if $y = f(x)$ and $x = g(t)$, then the derivative of y with respect to t is,

$$\frac{dy}{dt} = \frac{dy}{dx} \times \frac{dx}{dt}$$

Applying the chain rule to the right side of the equation,

$$\frac{\partial}{\partial a} S[x_a t] = \int_{t_1}^{t_2} \frac{\partial L(x, \dot{x}, t)}{\partial a} dt$$

Which equals,

$$\int_{t_1}^{t_2} \left(\frac{\partial L}{\partial x_a} \frac{\partial x_a}{\partial a} + \frac{\partial L}{\partial \dot{x_a}} \frac{\partial \dot{x_a}}{\partial a} \right) dt$$

Let's go back to see how we originally conceived of $x_a(t)$:

$$x_a(t) = x_0(t) + a\beta(t)$$

Where a is a number and $\beta(t)$ is a function whose endpoints at t_1 and t_2 yield $\beta(t_1) = \beta(t_2) = 0$.

Let's take the partial differential with respect to a

$$\frac{\partial x_a}{\partial a} = \beta$$

Differentiating again:

$$\frac{\partial \dot{x_a}}{\partial a} = \dot{\beta}$$

Substituting back into our prior equation:

$$\frac{\partial}{\partial a} S[x_a t] = \int_{t_1}^{t_2} \left(\frac{\partial L}{\partial x_a} \beta + \frac{\partial L}{\partial \dot{x_a}} \dot{\beta} \right) dt$$

Now it's time to integrate by parts (see section titled "Proving integration by parts").

Integration by parts states that:

$$\int u dv = uv - \int v du$$

Applying integration by parts to the far-right term in the integral, where $u = \frac{\partial L}{\partial \dot{x_a}}$ and $dv = \dot{\beta} dt$ we have:

$$\int \frac{\partial L}{\partial \dot{x_a}} \dot{\beta} dt = \frac{\partial L}{\partial \dot{x_a}} \beta - \int \left(\frac{d}{dt} \frac{\partial L}{\partial \dot{x_a}} \right) \beta dt$$

We'll substitute the right-hand side of the equation back into the third term of the prior expression for $\frac{\partial}{\partial a} S[x_a t]$. This gives us:

$$\frac{\partial}{\partial a} S[x_a t] = \int_{t_1}^{t_2} \left(\frac{\partial L}{\partial x_a} - \frac{d}{dt} \frac{\partial L}{\partial \dot{x_a}} \right) \beta dt + \left. \frac{\partial L}{\partial \dot{x_a}} \beta \right|_{t_1}^{t_2}$$

Now, all that remains is to evaluate the terms in the equation based upon our original goal (i.e., to find an $x_0(t)$ that yielded an Action derivative equal to zero) and the constraints that we set (i.e., $\beta_{t_1} = \beta_{t_2} = 0$).

By definition, for $x_0(t)$, the derivative of the Action is zero, so:

$$0 = \int_{t_1}^{t_2} \left(\frac{\partial L}{\partial x_a} - \frac{d}{dt}\frac{\partial L}{\partial \dot{x_a}} \right) \beta dt + \frac{\partial L}{\partial \dot{x_a}} \beta \bigg|_{t_1}^{t_2}$$

If $\beta_{t_1} = \beta_{t_2} = 0$, then $\frac{\partial L}{\partial \dot{x_a}} \beta|_{t_1}^{t_2} = 0$ and:

$$0 = \int_{t_1}^{t_2} \left(\frac{\partial L}{\partial x_a} - \frac{d}{dt}\frac{\partial L}{\partial \dot{x_a}} \right) \beta dt$$

But if the integral evaluates to zero, then the quantity within the parentheses must be zero or β must be zero. We previously stipulated that β could be any function satisfying $x_a(t) = x_0(t) + a\beta(t)$, so we know that *beta* is not zero. Therefore the quantity within the parentheses is zero, and we have:

$$\frac{d}{dt}\frac{\partial L}{\partial \dot{x_0}} = \frac{\partial L}{\partial x_0}$$

This is the Euler-Lagrange equation. Here is an oversimplified outline of how physicists employ the equation:

1. Compose a symmetry in terms of the Lagrangian equation (i.e., compose a Lagrangian that preserves an invariance under transformations of the Poincaré group).[26]
2. Compose the Euler-Lagrange equation for the Lagrangian.
3. Solve the Euler-Lagrange equation to produce a function that yields an Action derivative of zero.
4. Formulate the laws of motion and the symmetry of the system.

The Principle of Least Action holds that physical systems do not change the value of the Action. The symmetry transformation groups that obey the Principle of Least Action (i.e., whose transformations leave the Lagrangian stationary) are the symmetry groups that represent the physical laws of the universe.

[26]Named for Henri Poincaré (1854-1912) who was one of the first mathematical physicists to find the relevance of symmetries in physical laws under various transformations.

3.18 Proving Noether's theorem

"Que nous representent en effet les phenomenes naturels, si ce n'est une succession de transformations infinitesimales, dont les lois de l'univers sont les invariants?"
("What do natural phenomena represent other than a succession of infinitesimal transformations, whose invariants are the laws of universe?")
—Sophus Lie, in 1895

Put in simplest terms, Noether's theorem states that **wherever we have an invariance, we have a symmetry, and wherever we have a symmetry, there is a conservation law.** This straightforward theorem is key to our understanding of the fundamentals of the physical world.[27] Physicists interpret Noether's theorem as an expression of the symmetries of the Lagrangian (see prior section titled "Proving the Euler-Lagrange equation"). It permits us to conceptualize physical laws as symmetries, and to model our symmetry relations as Lagrangians that satisfy the Euler-Lagrange equation. Through these steps, physicists develop mathematical models of the operating system of the universe. Accordingly, every scientist should understand the meaning of Noether's theorem, and should have some notion of its proof. Unfortunately, a rigorous proof of Noether's theorem demands a bit more mathematics than we can feasibly provide herein. We present here a stripped-down mathematical proof of Noether's theorem that preserves the most important steps.

We need to prove that for each symmetry of the Lagrangian, there is a conserved quantity. This is equivalent to saying that when we change the coordinates of the system, then there is a quantity described by the Lagrangian that does not change. We shall need a solid understanding of the meaning of the Lagrangian, plus familiarity with the chain rule of differentiation (see section titled "Proving the chain rule").[28]

We can think of the Lagrangian as an expression of the energy state of a system as it moves incrementally through small changes in its coordinate system.

So, if q_i is one of n coordinates, a small change in the coordinate may be expressed as:

$$q_i \rightarrow q_i + \epsilon K_i(q)$$

[27] The theorem was published by Emmy Noether, in 1918, as "Invariante Variationsprobleme," in: Nachrichten von der Gesellschaft der Wissenschaften zu Gottingen, Mathematisch-Physikalische Klasse 1918:235-257, 1918.

[28] As in the prior proof of the Euler-Lagrange Equation, we loosely follow the steps described by David Morin in "Introduction to Classical Mechanics With Problems and Solutions", Cambridge University Press, Cambridge, UK, 2008.

Here, ϵ is a small number and $K_i(q)$ is a function that applies to the ith coordinate that may depend on the values of some or all of the total number of coordinates.

So, a change in coordinates for the Lagrangian will alter the Lagrangian (assuming the Lagrangian is a variant under a change of coordinates) as a function of all the changed coordinates and their first-order derivatives:

$$L = L(q_i, \dot{q}_i)$$

Because we stipulate that the Lagrangian does not change with a small increment in coordinates, we know that:

$$0 = \frac{\partial L}{\partial \epsilon} = \sum \frac{\partial L(q_i)}{\partial \epsilon} + \frac{\partial L(\dot{q}_i)}{\partial \epsilon}$$

The chain rule tells us that if $y = f(x)$ and $x = g(t)$, then the derivative of y with respect to t is:

$$\frac{dy}{dt} = \frac{dy}{dx} \times \frac{dx}{dt}$$

Applying the chain rule, we have:

$$0 = \frac{\partial L}{\partial \epsilon} = \sum \frac{\partial L}{\partial q_i} \frac{\partial q_i}{\partial \epsilon} + \frac{\partial L}{\partial \dot{q}_i} \frac{\partial \dot{q}_i}{\partial \epsilon}$$

But we know that:

$$q_i \to q_i + \epsilon K_i(q)$$

So,

$$\frac{\partial q_i}{\partial \epsilon} = K_i(q)$$

Substituting back into our equation,

$$0 = \frac{\partial L}{\partial \epsilon} = \sum \left(\frac{\partial L}{\partial q_i} K_i(q) + \frac{\partial L}{\partial \dot{q}_i} \dot{K}_i(q) \right)$$

Now we come to the point where we can apply the Euler-Lagrange equation. To review, when there is an invariance of the Lagrangian, the Euler-Lagrange equation specifies:

$$\frac{d}{dt} \frac{\partial L}{\partial \dot{x_0}} = \frac{\partial L}{\partial x_0}$$

In this case,

$$\frac{\partial L}{\partial q_i} = \frac{d}{dt} \frac{\partial L}{\partial \dot{q}_i}$$

Substituting back into our equation:

$$0 = \frac{\partial L}{\partial \epsilon} = \sum \left(\frac{d}{dt} \frac{\partial L}{\partial \dot{q}_i} K_i(q) + \frac{\partial L}{\partial \dot{q}_i} \dot{K}_i(q) \right)$$

But, $\dot{K}_i(q) = \frac{d}{dt} K_i$. Substituting back into our equation:

$$0 = \frac{\partial L}{\partial \epsilon} = \frac{d}{dt} \sum \left(\frac{\partial L}{\partial \dot{q}_i} K_i(q) \right)$$

Therefore, when we have a Lagrangian that does not change with incremental alterations in the coordinates, we ALWAYS have a quantity, $\frac{\partial L}{\partial \dot{q}_i} K_i(q)$ that does not change over time. A quantity that does not change over time is, by definition, a conserved quantity. **Therefore, we have proven that when there is a symmetry in the Lagrangian, there must be a conserved quantity. This is Noether's theorem.** [29]

[29] In the section titled "Proving the canonical commutation relation," we discussed the mathematical relationship between properties that happen to be the Fourier transforms of one another (i.e., paired Fourier conjugates). Noether's theorem extends to the Fourier conjugates in the following way. **When there is an invariance that applies to one of the paired conjugates, then the other conjugate will be conserved.**

3.19 Proving the electron field must contain a conserved quantity

The fundamental relationship between invariances, conservation laws, and symmetries are described mathematically as Noether's theorem, which we can paraphrase as: **Wherever we have an invariance, we have a symmetry, and wherever we have a symmetry, there is a conservation law.** Let's see how Noether's theorem applies to charge conservation.

- Electrons, like all fundamental particles, can be interpreted as wave packets.
- We have shown in a prior section that the phase of a wave cannot influence the outcome of any experiment.[30]
- As such, electron waves display a symmetry (i.e., no observable changes when their phase changes).
- We apply Noether's theorem to infer that the field must have a conserved quantity.
- We choose to call the conserved quantity "charge" (we could have named it anything we like).

The properties of charge derive from the equations that describe the electron field (not covered in this section). Symmetries of quantum fields wherein the phase can be adjusted without producing any observable change, are known as gauge symmetries.[31] Gauge symmetry accounts for the electromagnetic field theory and the law of conservation of charge.

[30]See section titled "Proving that a phase of a wave equation does not change the probability distribution for the wave equation".

[31]It may seem oxymoronic that a quantum theory based on the impossibility of measurement (i.e., no observable change with phase) would be called a gauge theory, inasmuch as the term "gauge" would seem to imply that the theory involves measurement. A seminal work leading to our current usage of the term "gauge theory" was written in German by Hermann Weyl (1920), who described a type of invariance requiring mathematical calibration ("eich Invarianz"), and the German phrase was translated into English as the misleading term, "gauge invariance."

4

Existences

> *"Science is a conceptual description and classification of our perceptions, a theory of symbols which economises thought. It is not an explanation of anything."*
>
> —Karl Pearson, *The Grammar of Science*, Preface to 2nd edition, 1899.

This is the chapter where physics meets metaphysics. In this chapter, a general proof is provided for precedence paradoxes, an informal proof is provided for Gödel's incompleteness theorem, and it is further shown that randomness is an unprovable property. It is proven herein that nothingness must have physical properties, and that mathematical symmetries provide a rational explanation for all of the observable properties of the universe. Answers, in the form of proof, are provided to such questions as: "How do we prove existence?", "How do we determine an object's fundamentality?", "What accounts for the apparent existence of spacetime?", "Is existence predetermined?", and "Why does existence exist?".

DOI: 10.1201/9781003516378-4

4.1 Proving that only the past is observable

"We must look for a long time before we can see".
—Henry David Thoreau

Because we exist in the present time, it seems logical to believe that we can observe the present time. This is not the case. All of our observations are based upon capturing light that has traveled to us at the moment of occurrence of some event. For example, if you were to stand in a room and suddenly noticed another person walking into the room, you might assume that you had observed an event occurring in the present time. This is obviously not the case because some finite amount of time must elapse while the light reflected by the individual's entrance travels across the room, to reach your eyes. More time is spent as the light that registers on your retinas is converted to a neural signal that is processed and interpreted by your brain. All in all, what you perceive is your mental reconstruction of an event that occurred in the recent past. There is no possible way by which events occurring in the present can be observed in the present.

We also cannot see into the future. Science fiction writers occasionally describe a future that is waiting patiently for us to arrive. Numerous stories have assumed that a clever inventor will construct a device that enables us to hop into the distant future and to hop back to the present, as the mood suits. A safer bet is to say that the future is something that does not exist, at present, and will not exist until some time in the future (which does not exist, at present). Niels Bohr is said to have quipped, "It's hard to make predictions, especially about the future." [1]

Seeing into the distant past is a common practice for astronomers. The image of a star located 13 billion light years from Earth must travel for 13 billion years before it reaches us. Therefore, astronomers who examine such a star are actually seeing what the star looked like 13 billion years ago, some hundreds of thousands of years after the Big Bang. For astronomers, seeing distant objects is equivalent to observing the distant past.

Though we cannot observe the present or the future, we can nonetheless extrapolate forward from the past and hazard a few assertions about the evolution of existence.

- Anything that can exist will exist, in a universe of infinite spacetime.

  If something can exist, then there is a non-zero probability that it will exist, and an event with a non-zero probability will occur when there is no limit

[1] This same quotation has been variously attributed to both Niels Bohr and to Markus H. Ronner.

to the wait time (i.e., if we can wait for eternity). This is a generalization of the adage known to all pessimists as Murphy's law: "Anything that can go wrong will go wrong."

- Anything that cannot exist now has not existed and shall never exist.

  This follows from time-shift invariance. One moment in time is equivalent to another, and if an outcome is impossible at this moment, then it is impossible at all other moments.

- New things emerge from old things. Put simply, there are no new things . . . there are just old things that are new to us.

  Everything we observe in the universe can be traced to events occurring with pre-existing objects. Every "new" atom of water forms from pre-existing Hydrogen and Oxygen. Every atom of Hydrogen and Oxygen arose from aggregations of protons and neutrons and electrons. Every proton arose from quarks and gluons. We can argue that the fundamental constituents of the universe were new at some time in the distant past, but they are certainly not new now.

  We can apply the same analysis to so-called "new" data. What we think of as new data is just old data that has been updated. Numbers that have no relation to prior values have no relevance to anything. It is only when data values change, in relation to pre-existing data, that we can find trends and discern meaning.

- Anything that comes into existence will persist until subsequent events end its existence.

  We see chemicals that break down, machines that cease to function, and living organisms that, as a rule of existence, must eventually die. We infer from this that everything in the universe will fade into non-existence over time. This is not true. The laws of conservation tell us that objects stay as they are until some event alters conditions and produce a change (i.e., things do not happen without a cause). The fundamental forces and particles are theoretically eternal unless terminated by some external event (such as annihilation via interaction with an anti-particle).

4.2 Proving the general solution to precedence paradoxes

The generalized precedence paradox is described as follows:

- Condition 1: A requires the pre-existence of B.
- Condition 2: B requires the pre-existence of A.
- Statement of Paradox: A and B cannot exist, but they do.

Because we can create A if we have B, and we can create B if we have A, the precedence paradox is sometimes stated as "Which comes first: A or B?"

The precedence paradox that everyone knows is the chicken and egg paradox. You cannot have a chicken until you first have an egg. You cannot have an egg unless you first have a chicken. Which came first, the chicken or the egg? The solution to the paradox is that during evolution, the animal kingdom was preceded by the evolution of single-celled organisms. In single-celled organisms, the egg and the organism are one and the same. Every animal, including chickens, can trace their evolutionary lineage to a point where the egg and the developed organism arising from the egg are unified. The answer to the question, "Which came first, the chicken or the egg?" is that the chicken and the egg were the same cells back in the time when the chicken's direct cellular ancestor was a single-celled eukaryote.

At this point, let's take a deep breath as we move away from the creation of living things, and we move toward the creation paradoxes that involve creations of the human mind; beginning with the computer. When you turn on your computer, the computer "boots up", the term being a shortened form for "bootstrapping." Every time you start your computer, a bootstrapping paradox must be solved. Here is the paradox:

1. Computer hardware requires a software operating system for any kind of functionality. Without software, a computer is just a paper weight.
2. Software operating systems cannot operate without functioning computer hardware. Without hardware, software is just a wish list.
3. When you turn on the power to your computer, the hardware cannot begin to function until it receives software instructions, but the software instructions cannot be accessed by the hardware until the hardware begins to function.
4. Hence, computers cannot exist.
5. But, computers do exist.

This human-created dilemma, which we unknowingly solve every time we turn on a computer, is referred to as a bootstrapping paradox. The term bootstrapping derives from an absurdist trope in which boys are instructed to pull themselves up by their bootstraps. Although it is certainly possible for a standing boy to pull his boots on with his bootstraps, it is impossible for a boy to gain a standing position by pulling the straps of a booted foot; no matter how hard he may pull. The term refers to a class of paradoxes in which some step in a process requires the completion of some earlier step, which itself requires the completion of a later step.

Here is how computers solve their bootstrapping paradox. At start-up, the operating system is non-functional. A few primitive instructions hardwired into the computer's processors are sufficient to call forth a somewhat more complex process from memory, and this newly activated process calls forth other processes, until the software operating system is fully up and running. The cascading re-birth of active processes takes time, and explains why booting your computer may seem to be an unnecessarily slow process. **Computer scientists solved the paradox by designing a tiny start-up kernel wherein both the hardware and the software are one and the same.**

Whenever you encounter a bootstrapping paradox, try to rephrase it as a precedence paradox. After doing so, see if you can't find a solution based on a duality origin, as follows. **If A and B were equivalent, at the origin of their development, then the paradox disappears.**

There are numerous examples of precedence paradoxes that were solved by establishing the equivalence (at their origins) of items that each depend on the pre-existence of one another. We can demonstrate the point with a few examples.

- The enzyme and enzyme synthesis paradox

  To build a protein, our cells require a protein-synthesizing machinery, which is itself made of protein. The paradox here is that the product of the protein-synthesizing machinery (i.e., the protein) must exist prior to the existence of the protein-synthesizing machinery (which is made of proteins).

  To find the solution to this paradox, we must understand that all metabolism involves catalysts that facilitate reactions by holding the substrates of a reaction in close proximity to one another so that they can interact to form a product that is released inside the cell. Proteins make excellent catalysts because they can evolve as structures that fit just about any substrates. Proteins that serve as cellular catalysts are known as enzymes. Other molecules (e.g., surfaces of rocks, minerals, RNA molecules) could also serve as catalysts, but only if they happen to have the requisite physico-chemical features.

Ribosomes are highly specialized molecules that do the bulk of work involved in translating RNA molecules into new proteins. Ribosomes contain proteins and ribozymes, catalytic RNA molecules that facilitate the translation of messenger RNA into amino acid chains (i.e., proteins). As it happens, ribozymes are also capable of catalyzing their own synthesis. Under laboratory conditions, ribozymes can sequentially attach dozens of nucleotides to primer sequences, without the help of enzymatic proteins.[2] We can imagine that the very earliest metabolic machinery in cells may have consisted of a molecule that participated in the production of encoded amino acid chains (i.e., primordial proteins) and copies of itself (RNA chains carrying the code). Paradox solved! One species of molecule had the primitive features of both protein and protein-synthesizing machinery. It was only after some period of evolution that the two features became distinguishable and represented by different sets of molecules.

- The virus and cellular host paradox

We know that viruses and cells are both ancient forms of replicating life. Viruses infect and replicate within host cells (such as bacteria, plant cells or animal cells) and do not replicate outside of host cells. Thus, viruses require the pre-existence of living cells. Living cells evolved the cellular machinery for replication (i.e., mitosis and meiosis) over many millions of years, existing in a form incapable of replication in the interim. During this early period, cells could only have increased in number by distributing their genetic material directly by the movement of genetic fragments from one protocellular compartment to another. The traversal of DNA from one host cell to another is a defining feature of viruses. Therefore, cellular life requires the pre-existence of viruses, much as viruses require the pre-existence of cellular life; and this presents a paradox. The paradox is solved by asserting that the virus and host cell were unified when life began on Earth. Primordial cells were compartmentalized packages of nascent metabolic machinery, containing genetic templates that could replicate to some extent, producing sequences that occasionally moved from one cell to another. Eventually, cells and viruses evolved as distinguishable organisms.

The unified origin of viruses and host cells is just one of many proposals addressing the origins of life on Earth. Nobody really knows how cellular life began, with any degree of certainty. But there are credible solutions for the precedence paradoxes that biologists have encountered.

- Electricity and magnetism paradox

[2]Interested readers may wish to examine the following manuscripts: Johnston WK, Unrau PJ, Lawrence MS, Glasner ME, Bartel DP. RNA-catalyzed RNA polymerization: accurate and general RNA-templated primer extension. Science 292:1319-1325, 2001; Zaher HS, Unrau PJ. Selection of an improved RNA polymerase ribozyme with superior extension and fidelity. RNA 13:1017-1026, 2007; Wochner A, Attwater J, Coulson A, Holliger P. Ribozyme-catalyzed transcription of an active ribozyme. Science 332:209-212, 2011.

Electric waves are generated by magnetic waves, and magnetic waves are generated by electric waves. Because each requires the pre-existence of the other, we can infer that there must be an electromagnetic wave, with properties of both.

- Class and Classification paradox

  A classification is a collection of classes. The members of a class all share certain properties (so-called class properties). There are also well-defined relationships among all the different classes. The classification of all organisms, living and extinct, is the best-known example of its type; but everything in the universe, living or dead, material or abstract, would be better understood if it were included in a well-constructed classification.[3]

  A battle that has been fought for decades, within the biomedical community, over the method by which biological classifications can be built. The two opposing camps consist of the computational bioinformaticians and the classical taxonomists.[4,5] Computational bioinformaticians believe that the members of a biological class will exhibit similar genes and traits, and that these similarities can be sorted into classes of species, without imposing any biases and preconceptions. Taxonomists believe that a classification cannot be built upon similarities among species, insofar that similarities do not reveal the fundamental relationships among the members of a class or among one class and another class. For example, terriers and house cats share many similarities: similar size and weight, four paws and a tail, fur, and so on. Nonetheless, dogs and cats belong to different classes of animals and have a different ancestral lineage. Taxonomists argue that it is wrong to classify dogs and cats or any organisms based upon the similarities that they share. Hence, algorithms cannot create a useful classification.

  Taxonomists create classifications of organisms based on their personal worldviews in which various organisms have preconceived relationships with other organisms. Hence, the taxonomist's worldview contains a formed conception of the classification of things; which presupposes that the classification already exists as an abstraction. The computational bioinformaticians argue that taxonomists cannot build a classification without first having the classification in their mind, thus biasing the outcome. Hence classical, taxonomy cannot be relied upon to build a useful classification.

---

[3] If interested, you might try reading the author's book titled *Classification Made Relevant: How Scientists Build and Use Classifcations and Ontologies*, Academic Press, 2022.

[4] Here, the word "taxonomist" applies to any biologist who contributes to the classification of all living organisms. Classical taxonomists add species to a hierarchy of pre-existing classes based on relating the properties of the species to the properties of a specific class.

[5] Computational bioinformaticians use algorithms that process a collection of biological properties and automatically organize the properties into an interconnected collection of classes into which species can be assigned, based upon shared similarities.

The classical taxonomist must overcome a bootstrapping paradox. Namely, it is impossible to build a classification without a pre-existing collection of related classes in which to insert class members. But, it is impossible to have a pre-existing collection of related classes without first having the built classification that contains the classes and their relationships.

In practice, the classical taxonomist begins with a root object that embodies the fundamental features of every class and every species within the classification. In the case of the taxonomy of living organisms, this root object might be "the living cell". Once the root object exists, the taxonomist can begin to create broad subclasses containing properties that are inclusive of the class and exclusive of other classes. Then, based upon observing the properties of the derived classes, she might define additional classes that include some organisms and exclude others. This goes on until every organism has an assigned class, and every class is a subclass of a parent class, in a lineage that extends backward to the root class. **The root class, which is the overall superclass, contains every member of the classification and embodies the full classification. That is to say, the very first class of the classification establishes the premise upon which the full classification emerges.** In summary, the classification and its root structure are one and the same object, when work on the classification begins.

For theoretical physicists, a similar classification paradox holds. To build a classification, we need to know the root of the symmetries from which all physical reality has emerged. Our classification of the forces of nature (encompassed by the theory of gravitation and the Standard Model) has all emerged from observed symmetries, but at this point, we do not have a root symmetry that accounts for the emergence of our observed symmetry classes.[6] We cannot have a classification of the symmetries of nature without a classification of the forces of nature that arise from symmetries. [7] But we cannot classify the forces of nature without the unifying root symmetry that preceded the observable forces of nature (i.e., we lack a unifying theory of the forces of nature). This root symmetry has not, as yet, been found.[8] We will continue our discussion of the root of the forces of nature in section titled "Proving all fields derived from one root field".

- Time and space paradox

We cannot experience the passage of time unless something happens, and everything that happens involves movement through space. We cannot

---

[6]For a discussion of the relationship between symmetries and physical laws, see section titled "Proving the operating system of the universe is determined by symmetries".

[7]The symmetries of nature are currently classified as various Lie groups. Theoretical physicists are seeking a single group that encompasses every aspect of physical reality.

[8]For an excellent discussion of the possible symmetry classifications of quantum physics, see manuscript written by Robert Arnott Wilson titled "A group-theorist's perspective on symmetry groups in physics", Arxiv Math.GR 2009.14613v5, December 20, 2020.

experience space without time separating locations in space. Otherwise, objects could move to any place in space instantaneously, and neither distance nor location would have meaning. The solution to the precedence paradox is that space and time are unified as spacetime.

The solution to any precedence paradox requires that the interdependent conditions must have been one and the same thing at the point of their origin.

When this general solution cannot be applied, the paradox is truly unsolvable, and the conditions of the paradox cannot exist. Science fiction enthusiasts will certainly be familiar with the precedence paradox of the homicidal time traveler. Here, a murderous daughter travels back in time to visit her father in his boyhood. Whereupon, she slits her father's throat. By doing so, the father cannot sire the daughter, and the daughter is not born. Because the daughter cannot exist, she cannot travel into the past to murder the father. The general solution to the precedence paradox does not help in this situation. We must conclude that the conditions of the paradox are impossible (i.e., time travel, as described here, cannot happen).

4.3 Proving all fields derived from one root field

"But – once I bent to taste an upland spring
And, bending, heard it whisper of its Sea".
—Ecclesiastes

One of the most sought-after solutions in modern physics is the unification of the four forces of nature (i.e., the electromagnetic force, the weak force, the strong force, and gravity) under one consistent theoretical model. [9] Current thinking posits that at the earliest moment of the Big Bang (i.e., in the first 10^{-43} seconds), all of the forces (and their fields) must have been unified as a single field that soon thereafter split into the individual fields that we observe today. This theory is based in part on calculations stemming from the Standard Model indicating that the energies existing in the first 10^{-43} seconds of the Big Bang are so intense that the fields, as we know them, could not exist. The assumption is that there was a primordial force out of which our observable forces of nature soon emerged. There is also the belief, based upon the general trend of scientific advancement, that the operating principles of nature are simple, and the role of science is to find simpler and simpler ways of encapsulating the laws of nature. Basically, one unifying field is simpler than many fields having independent origins and operational principles. As yet, physicists have not found a model that unifies all of the forces of nature plus gravity under a single model that fully accounts for our observed reality.

At this point, we might build an argument that bolsters support for a unified field theory.

- The properties of one field are dependent upon the properties of another field.

  Put another way, the properties of a field depend upon terms included in the equations of motion of other fields.[10] In particular, different fields operate under the same set of fundamental constants.

[9] The Standard Model, which deals with three of the known fields of nature, but which omits gravitation, is considered to be our most accurate model of physical reality and has proven to be an excellent predictor of experimentally verifiable quantum events. Nonetheless, it has its limitations. The Standard Model fails to provide a candidate particle for dark matter. It does not explain baryon asymmetry (i.e., more protons than anti-protons in the universe). It fails to provide a satisfactory model for quantum gravity and thus does not unify the forces of nature. Sad to say, not every prediction coming forth from the Standard Model is true. A more complete discussion of the limitations of the Standard Model can be found in the Particle Data Group's Review of Particle Physics, in the journal Chinese Physics C.40, 2016, page 290.

[10] Using terminology taken from quantum field theory, terms in the Lagrangian or the Hamiltonian depend on several fields.

- All particles, regardless of their field of origin, have movement and can exchange momentum with other particles from their own field or from other fields.

  The universality of energy, movement, and momentum confirms that the fundamental properties of the different fields are equivalent. It is worth noting that massive particles of a particular field do not directly interact with one another. In the case of electron-electron interactions, their like charges would repel, preventing direct collisions. In the case of other massive particles, their mutual gravitational attraction would reach infinity as the distance between particles falls to zero. Interactions between particles of one field are mediated by virtual particles of another field.[11] Because field interactions are mediated by other fields, we can see that the properties of all fields are coordinated under mutual laws.

- The particles of a field may derive from the particles of other fields.

  For example, τ particles can decay into fundamental particles of different fields (electrons, neutrinos of different types, quarks, pions).

- The first three items satisfy the conditions of a precedence paradox. For two fields (A and B), field A cannot exist without the pre-existence of Field B. Likewise, field B cannot exist without the pre-existence of Field A. The general solution to any precedence paradox is that interdependent conditions must have been one and the same thing at the point of their origin (see section titled "Proving the general solution to precedence paradoxes"). In the case of the forces of nature, the solution of the precedence paradox is that all of the fields and their forces must have been one and the same, at some point in time.

- When we suggest that all of the observed fields in our universe came from a single field, we are equivalently suggesting that all of the symmetries of the universe derived from a single symmetry. We also know that symmetries may emerge from symmetries. Hence, the notion that our observable symmetries are derived from a single symmetry seems plausible.

  As we shall discover in later sections, the laws of fields, and their forces, result from symmetries. Symmetries are subject to perturbations due to spontaneously occurring energy fluctuations that provide sufficient energy to "break" the existing symmetry rules for a short moment. This is the principle that made quantum tunneling microscopy a reality. When the symmetry rules are broken, it becomes possible, in theory, for a new and

[11] Virtual particles are produced by small fluctuations of field energy. These fluctuations occur because a steady zero-energy state is prohibited by the Heisenberg Uncertainty Principle. Specifically, if the energy at any location were always zero, then we would be certain of the energy level, thus violating uncertainty. Virtual particles are short-lived and unobservable.

stable symmetry to emerge.[12,13] Because one symmetry can spawn additional symmetries, it seems reasonable to imagine that the symmetries responsible for the Standard Model and gravity may have all begun with one symmetry. If that were the case, when we account for the one symmetry from which all other symmetries emerged, we will have succeeded in explaining the fundamental nature of existence.

Of course, these points do not constitute proof, but they furnish optimism that a viable unified field theory may emerge.

[12] For a detailed discussion of this subject, refer to: Dimakis N. Generation of new symmetries from explicit symmetry breaking. arXiv:2112.02776v1, Dec. 6, 2021.

[13] Antonio Garcia-Bellido has suggested that the evolution of species occurs through a similar process, writing "The evolution across several levels of complexity from atoms to molecules, to cells, to organs, and to organisms is an iterated symmetry break." See his manuscript, titled "Symmetries throughout organic evolution", in Proc Natl Acad Sci USA 93:14229-14232, 1996.

4.4 Proving fundamentality is, as yet, unprovable

"Couldn't Prove Had to Promise"
—Title of a poetry collection by Wyatt Prunty

We think that something is fundamental if it has no parts (i.e., is not composed of other things). On a very simplistic level, we can review some familiar objects and apply our definition to determine the object's fundamentality.

- Mobile phones. Certainly, mobile phones are not fundamental. They are composed of transistors, circuits, batteries, and innumerable parts.
- Iron ore. Certainly iron ore is not fundamental. It is composed of oxidized iron and aggregate rock.
- Iron atoms. Certainly, atoms of iron (or of any other element) are not fundamental. Each atom is composed of a nucleus and electrons.
- Nucleus of an iron atom. Certainly, the nucleus of iron (or of any other atomic nucleus) is not fundamental. It is composed of neutrons and protons.
- Protons. Certainly, protons (and neutrons) are not fundamental. They are composed of quarks and gluons.
- Electrons. Yes, electrons seem to be fundamental. They are point particles, having no size, and containing nothing.
- Photons. Yes, photons seem to be fundamental. They are point particles, having no size, and containing nothing.

We could continue on if we wished. For physicists, the fundamental particles predicted from the Standard Model are[14]

- Six flavors of quarks (up, down, strange, charm, bottom, and top).
- Six types of leptons (electron, electron neutrino, muon, muon neutrino, tau, tau neutrino).
- Twelve gauge bosons (force carriers): the photon of electromagnetism, the three W and Z bosons of the weak force, and the eight gluons of the strong force.
- The Higgs boson.

All of these particles are point particles, having no size, consisting entirely of their properties, as described in the Standard Model of physics. Together

[14] All of which have now been shown, experimentally, to exist.

with gravity (which is not included in the Standard Model), we have all of the fundamentals of the observable universe. Or do we?

String theory emerged in the late 1960s as a way to replace all of the known fundamental particles with one fundamental object known as a "string". [15] The string is hypothesized to be a very small [16], unobservable, vibrating loop. One string may manifest as any of the point particles predicted by the Standard Model, depending upon how it happens to oscillate. In this scenario, the "string" is the only fundamental particle in our universe. Everything we see in our universe is composed, in one way or another, from the fundamental string. The 24 "fundamental" particles of the Standard Model are now seen as non-fundamental emergents of the single fundamental "string". As it happens, string theory is unproven and does not qualify as the new essence of reality.[17]

Putting all considerations of string theory aside, we can use our current understanding of quantum theory to directly challenge the notion that fundamentality is determined by the physical composition of an object (i.e., when something is not composed of other things, it is considered fundamental). If the so-called fundamental particles have properties determined by their respective fields (i.e., quarks have properties of the quark field and bosons have properties of the boson field), then particles are composed of properties and are therefore not fundamental. In the case of every so-called fundamental particle, their respective field properties are "more fundamental."

Now, let's imagine the situation wherein a putative fundamental property of a field came into existence as a consequence of some pre-condition. Put another way, some property X cannot exist unless some other property Y exists. In this case, property X is not fundamental insofar as its property is a consequence of some other property that would be "more fundamental".[18]

In quantum field theory, we run into a situation where a field, and its properties, are dependent upon some other condition for its existence. For example, the magnetic field requires a magnetic vector potential (not vice versa). Likewise, the electric field is dependent upon the electric scalar potential (and not vice versa). In both cases, there is something more fundamental than the

[15]Gabriele Veneziano is credited with writing the first manuscript on string theory, in: "Construction of a crossing-simmetric, Regge-behaved amplitude for linearly rising trajectories." Il Nuovo Cimento A 57:190-197, 1968.

[16]The string's length is equal to one Planck length, or 1.6×10^{-35} meters

[17]String theory has become a disappointment to many theoretical physicists as it does not provide answers to the questions it was intended to address, namely, the derivations of the cosmological constant, the fine structure constant, the gravitational constant, and the Planck constant. Furthermore, string theory has led physicists to predict the existence of particles that have not been shown to exist. A growing body of physicists seems to have lost hope in this once-promising avenue of research.

[18]A non-fundamental property would be a property that emerges from other, more fundamental properties. In a sense, a property composed of other properties is not fundamental.

field. Therefore, the field, its properties, and its quantum particles, are all non-fundamental. Similarly, we can think of the so-called fundamental constants of the universe as dimensionless properties that derived from a precondition that escapes our grasp (see section titled "Proving all fundamental constants are dimensionless").

There is reason to suspect that everything we infer about our universe has emerged from fundamental(s) that are unknown to us, and quite possibly unobservable. If we give our suspicions any credence, we might infer that our evaluation of fundamentality is weak. We may be including non-fundamentals as fundamentals, and excluding fundamentals as non-fundamentals. Put simply, we may be fundamentally wrong about fundamentals. In any event, we infer that fundamentality is a condition that we are not presently equipped to establish through proof.

4.5 Proving the universe operates according to universal laws

"Welcome to reality, where everything sucks and your feelings don't matter".
—Drew Carey

The evidence that the universe operates by universal laws, and that nothing in the universe occurs in a lawless way is provided by three casual observations:

- Repeatability

  Any experiment performed on Monday will produce the same results when repeated on Tuesday; without exception. Unencumbered by laws, a single cause could result in different effects, at different times or in different places.

  As a corollary, the repeatability of nature allows us to predict the specific outcome of a causal event. We can predict with confidence that the outcome of a specific cause will produce the same effect tomorrow as it produced today. By examining the predictable relationships between cause and effect, we can determine the law or laws involved.

- Stability

  In the absence of universal laws, the events occurring in the universe would be unpredictable and chaotic. There would be no formation of stable atoms, molecules, matter, or living organisms. Events that occur at one moment in a lawless universe might never occur again.

- Sameness

  Wherever we look in the night sky, we see a similar landscape, consisting almost entirely of galaxies and their contained stars and planets. Nearly all the galaxies have about the same appearance: whirling flat disks with a central bulge. The light emanating from one star looks more or less like the light emanating from any other star (adjusting for red-shift). The atoms found in the interstellar void are the same everywhere (i.e., mostly Hydrogen, with a bit of Helium), and the atoms found in and around stars are limited to the same 94 naturally occurring elements, everywhere.[19] The cosmic microwave background is more or less uniform in every direction.

[19]As an example of "sameness" in action, all supernovae are thought to arise from the same set of conditions, yielding supernova of equivalent brightness. Astronomers apply the sameness of supernovae to the concept of "standard candles." If we know the distance from Earth to any one supernova, we can compute the distance to any other supernova, by applying the inverse square law to the measured brightness of the different supernova, as registered here on Earth.

> The sameness of the night indicates that the laws of nature are the same everywhere.

In other sections of this book, we will see that there is a limited assortment of conservation laws, each indicating a particular physical invariance. These invariances are best understood as symmetries, and these symmetries account for the operating system of reality.

4.6 Proving measurements have no absolute meaning

"Only theory can tell us what to measure and how to interpret it".
—Albert Einstein

We measure things all the time, and we expect our measurements to be meaningful, precise, and valid. But are they ever?

- Measurement is self-referential and cannot be accepted as fact.

  Our measurements are made in units, the meaning of which is determined by our measurements of the unit. Hence, every measurement is dependent upon some preceding measurement. There is no fundamental measurement (i.e., a measurement that is independent from a previous measurement) that is known to be true and from which which we can build a logical chain of truth.

- The basic units of spacetime geometry have no absolute meaning.

  The theory of special relativity informs us that space and time are relative quantities depending on frames of reference. The speed of light is observed to be constant in different frames of reference, but the speed of light is determined with yardsticks and clocks, whose measurements will vary in different reference frames. Our measurements of the speed of light have told us a great deal about our measuring devices (i.e., yardsticks and clocks) but nothing much new about the speed of light.[20]

  Likewise, the theory of general relativity informs us that there is no way to shield our measuring devices from their gravitational field, which slows the speed of clocks. Therefore our measurements of constants will always depend upon the gravitation field in which the measurement is conducted. A constant measured on Jupiter may have a different value than the same constant on Uranus.

- All measurements have limits to their precision.

  Even if we had measurements whose meaning is understood, our ability to measure anything with exactitude is limited. This assertion follows directly from the Heisenberg Uncertainty Principle (see section titled "Proving the Heisenberg uncertainty principle"). But we can construct a simple and intuitive argument based upon the wave nature of matter.

[20] Without measuring the speed of light, we can prove that the speed of light is constant in a vacuum. But knowing that the speed of light is constant does not tell us anything about its measure.

1. Everything physical has wave properties.
2. None of the parameters of waves can be measured with total precision insofar as waves are spread in space and time.
3. Therefore, nothing of a physical nature can be precisely measured.[21]

- We must understand how our measurement influences that which we are seeking to measure.

  In brief, the phenomenon known as the collapse of the wave function tells us that fundamental particles behave like wave packets until they are detected by a sensor, at which time they begin to behave like particles. We discuss this topic in greater depth in the section titled "Proving that the collapse of the wave function invalidates measurement."

- We never fully understand the subjects of our measurements.

  Try as we may, we do not understand fundamental physical systems. Simple concepts such as entropy, randomness, determinism, and causality, all lie just beyond our grasp. Our most widely accepted notions of space, time, energy, and forces are built upon theory. Hence we can never be confident of the meaning of measurements when the subjects of our measurement have no absolute meaning.

Physics is the study of physical relationships. The job of the physicist involves translating relationships (that are based upon observation and measurement) into formulae. In a very real sense, physicists do not understand, or even attempt to understand, the meaning of measurements. They are concerned almost exclusively with describing the relationships among particles and forces, and the consequences of interactions, based on the formulae that express those relationships. Don't waste your time asking a physicist for the meaning of magnetism, charge, or the Big Bang. "Meaning" has no meaning to the work-a-day physicist. Questions of "meaning" are left to philosophers.

[21]It is quite amazing that in the absence of adequate measurements, scientists do a fairly good job at analyzing waveforms. In a manuscript by Emmanuel Candes, Justin Romberg, and Terence Tao, titled "Stable Signal Recovery from Incomplete and Inaccurate Measurements" (published in arXiv:math/0503066v2, math.NA, Dec 7, 2005), the authors showed the feasibility of reconstructing signals from sparse and inaccurate signal samplings.

4.7 Proving all fundamental constants are dimensionless

"Infinitely sharp constants are failed clocks. They imply complete delocalization in time".
—João Magueijo [22]

It is easy to show that fundamental constants must be dimensionless.

1. By definition, a constant is fundamental if it cannot be calculated in terms of more fundamental constants.
2. But a dimensional constant is, by definition, a constant that is defined and measured in terms of constant units.

   We know that whatever units are chosen to measure constants must themselves be constants. Otherwise, their value would change and the value of the dimensional constant would also change.
3. Therefore, a dimensional constant is not fundamental (**because** it is calculated in terms of other constants).

   Dimensional constants, such as the speed of light, are best interpreted as conversion tools. If the speed of light is measured as 3×10^8 meters per second, then we could convert a second to 3×10^8 meters per speed of light, and we could convert a meter to the speed of light times a second divided by 3×10^8. Doing so may seem silly, but astronomers measure distance in terms of light-years without the slightest compunction. All of our dimensional constants devolve into measured quantities represented by a chosen set of units that inter-convert with our constants.
4. Therefore, all fundamental constants must be dimensionless.

An example of a dimensionless constant is α, the fine structure constant, known to be approximately equal to $1/137$.[23] There are just a few dozen

[22] from Magueijo J. Connection between cosmological time and the constants of Nature. arXiv:2110.05920v3, March 30, 2022

[23] As discussed earlier, the fine structure of the Hydrogen atom's emission spectrum was observed to produce two narrowly separated lines, where only one line was predicted by the Bohr theory of the atom. The fine structural separation of spectral lines was quantified by Sommerfeld as a dimensionless value, $1/137$, and has been called the fine structure constant or Summerfeld's constant (1916). Since then, numerous interpretations for the physical meaning of the fine structure constant have been offered, and values for the constant have been more precisely established from experimental measurements and predictions based on quantum electrodynamics theory.

generally recognized fundamental dimensionless constants, most of which pop out of the equations of the Standard Model.

If fundamental constants are dimensionless, as they must be, then we are drawn into a set of very difficult questions, whose answers are legitimate topics of concern in mathematics, physics, and metaphysics.[24]

- The laws of our universe are determined by numbers, which are themselves numeric abstractions. How can an abstraction be a part of physical reality?
- Where do the dimensionless constants come from? That is to say, what accounts for the existence of dimensionless constants?

  For the most part, these constants pop out of the standard model. These are:

  - The 9 Yukawa coefficients for quarks and leptons
  - The 2 Higgs coupling constants
  - The 3 angles and a phase of the CKM matrix
  - The phase for the Quantum ChromoDynamics (QCD) vacuum
  - The 3 coupling constants for the gauge group $SU(3) \times SU(2) \times U(1)$
  - The cosmological parameter
  - The 7 or 8 constants for massive neutrinos

  The Standard Model, in turn, is developed from symmetries. Hence, dimensionless constants arise from the symmetries that lie at the heart of reality.

- How can we be certain that the dimensionless constants never change?

  Simply insisting that a constant is "constant" (i.e., unvarying in time and space) is not enough. Physical constancy needs to be proven.[25,26]

  At present, no convincing proof of the constancy of the dimensionless constants, over time, is available. The obstacles to proof come in two forms:

---

[24] For an additional discussion of dimensionless constants, see: Duff MJ. How fundamental are fundamental constants? arXiv 2:1412.2040, Dec 17, 2014 and Weinberg S. Overview of theoretical prospects for understanding the values of fundamental constants. Phil. Trans. R. Soc. London A310:249-252, 1983.

[25] For a discussion of the variation of dimensionless physical numbers on cosmologic scale, see Narimani A, Moss A, Scott D. Dimensionless cosmology. arXiv:1109.0492v4, Jun 26, 2012.

[26] If dimensionless constants vary in time and space, is there a dimensionless constant that describes the variation of constants? This enigmatic question is discussed in a manuscript by Jean-Philippe Uzan titled "Varying constants, gravitation and the cosmology", in Living Reviews in Relativity 14:2, 2011.

- We can try our best to make precise measurements of the value of the dimensionless constants, at different times, for comparison.[27] But our measurement equipment will never be perfect, and any small variations found in our results will be the subject to criticism.

- We can tie the value of a constant to some conserved quantity (energy or charge or momentum), but even our trusted conserved quantities may have some variation over large periods of time, on a cosmologic scale.

  For example, nearly all of the total observable energy in our universe comes from the cosmic background radiation. But we know that over the billions of years since the Big Bang, the frequency of the cosmic microwave background has decreased, and with the decrease in frequency, there has been a corresponding decrease in its total contained energy. This would suggest that energy is not conserved on the cosmologic scale. This being the case, it is not unthinkable that physical constants (dimensionless or dimensional) may also change when observed over very large periods of time.

- Do the dimensionless constants have an existence beyond time and space?

  Were the dimensionless constants created at the time of the Big Bang, or have they always existed, as a condition of reality? There really is not much reason to assume that our fundamental constants emerged from the Big Bang. Neither is there any proof that the symmetries of the universe arose as a consequence of the Big Bang. Speculation on the existence of so-called "pre-space" constants is a valid subject for exploration.[28]

- Are there dimensionless constants that are not fundamental?

  The fundamental forces of the Standard Model are all described by fundamental constants. Therefore, constants (dimensional or dimensionless) arising from the interplay of fields (and their forces) are derived from the fundamental constants and are therefore not fundamental. There are several ways of producing non-fundamental dimensionless constants.

  - By expressing the constant as a counted number. For example, Avogadro's number (about 6×10^{23}) is a dimensionless constant, though it is derived from dimensional measurements (see the section titled "Proving the atomic particle theory"), and is not fundamental.

---

[27]The most precise measurement achieved to date has been in the realm of atomic clocks, where the accuracy of 18 decimal places has been claimed. For details see Bloom BJ, Nicholson TL, Williams JR, et al. An optical lattice clock with accuracy and stability at the 10–18 level. Nature 506:71-75, 2014.

[28]For further discussion of this topic, see Tejinder P. Singh's monograph "Quantum theory without classical time: Octonions, and a theoretical derivation of the Fine Structure Constant 1/137", in the Intl J of Modern Physics D 30:2142010, 2021.

- By expressing constants as formulae whose dimensional parameters appear as ratios. The dimensions cancel out, to produce a dimensionless constant, but the dimensionless constant is derived from other quantities.

4.8 Proving constants of mathematics have physicality

"All things are number".
—Pythagoras of Samos (570 - 495 B.C.E.)

In the prior section, we argued that all fundamental constants are dimensionless.[29] Here, we argue that mathematical constants (such as π and e) qualify as constants of nature if they appear in the equations that describe natural laws. Here, we define a mathematical constant as a number whose value is defined by a mathematical operation. For example, π is the ratio of the circumference of a circle to its diameter, and e is the base of natural logarithms.[30] Dimensionless constants π and e, and a few other values that we shall soon mention, are not derived from other constants, and they are included in the equations that describe the natural laws that apply universally. We assert that such constants are fundamental constants of nature.

There is no accepted comprehensive list of the natural (i.e., physical) fundamental constants, but here are a few examples of numbers that seem to qualify:

- $1, 2, 1/2, 1/3, 2/3$ are all particle spin numbers.
- π, the ratio of a circle's circumference to its diameter
- $\sqrt{2}$, known as the Pythagoras constant, the positive root of $x^2 = 2$, and appearing in the basis for the SU(2) generators
- i, known as the imaginary unit, the root of $x^2 = -1$
- e, known as the Euler number, and evaluated as $\lim_{n \to \infty} \left(1 + \frac{1}{n}\right)^n$
- θ known as the golden mean, and defined as the solution to the equation $\theta = 1 + \frac{1}{\theta}$. The golden mean is approximately 1.6180339887.
- ϕ_m, known as the magic angle, a root of the second-order Legendre polynomial, applied widely in nuclear magnetic resonance spectroscopy, and evaluated as $\arctan \sqrt{2}$

The list goes on and on. An intriguing candidate for our list of the natural mathematical constants is the number zero, which indicates the absence of quantity. We all think we have nothing to learn about "zero", but some of the finest mathematical minds throughout history have failed to grasp its full

[29]See section titled "Proving all fundamental constants are dimensionless".

[30]The transcendental mathematical constants, such as π, are calculated, not measured. π has been calculated to over 62 trillion digits beyond the decimal point, a recent achievement credited to a team working from the University of Applied Science of the Grisnos, in Chur, Switzerland, in 2021.

significance. Just to get an idea of how zero vexes numeric reasoning, let's look at the following "proof" that $5 = 1$.

1. $5 = 5$ Obvious
2. $0/0 = 1$ Any number divided by itself is equal to one.
3. $5 \times 1 = 5 \times (0/0)$ Substituting 0/0 for 1
4. $5 = (5 \times 0)/0$ Distributing the numerator
5. $5 = 0/0$ Evaluating the numerator
6. $\therefore 5 = 1$

This proof, such as it is, can be applied to show that every integer is equal to 1. Our mistake, of course, was to assume that zero has a multiplicative inverse (i.e., an integer which, when multiplied by itself, yields the identity integer, 1). We know, of course, that 0/0 is undefined and is certainly not equal to 1. Because zero has no inverse under multiplication, its inclusion in the set of integers tells us that the set of integers under multiplication cannot be a group.[31]

For thousands of years, ancient mathematicians did not know quite what to do with zero. For a long time, the zero (or its ancient equivalents) was used as a placekeeper signifying an empty space in a number, without taking an active role in calculations. Historians seem to concur that the use of zero as a true number was not known to Indian mathematicians until the ninth century AD, preceded by several centuries by the Mayans . Western mathematicians did not seem to catch on until about the 13th century [32] To this day, nobody knows how to handle division by zero.[33] When our computers run up against division by zero, they typically scold us with a curt error message indicating that division by zero is forbidden. It is as though we humans are encouraged to enjoy all the benefits of mathematics save for one supreme commandment outlawing division by zero, the forbidden fruit of numerology.

What then, is the proper role of zero? Zero is the additive inverse, meaning that it the the number that results when any number is added to its inverse. Therefore, zero fits the definition of a mathematical constant. In addition,

---

[31]Remember that a defining property of a group requires that **every** member of the group must have an inverse that is also a member of the group.

[32]See JJ O'Connor and EF Robertson's essay, "A history of zero." in MacTutor November, 2000.

[33]When we divide any number, n, by smaller and smaller numbers, the result grows larger and larger. It would seem that the limit achieved when any number is divided by zero would be infinity. But if this were the case, and $n/0 = \infty$, then $n = \infty \times 0$, and this would apply to all numbers n. Then every number would equal $\infty \times 0$, and we might as well just chuck out all of mathematics and try our hands at writing poetry for a living.

we can list reasons why zero is a fundamental physical constant, with just a much significance as the fine structure constant, π or e.

- Zero is constant (unchanging over time and space).
- Zero is dimensionless (like all valid physical constants).
- Zero plays a central role in every conservation law insofar as the net change over time of all conserved quantities must equal zero.
- When we derive equations that govern physical processes, we encounter the zero (e.g., absolute zero temperature, zero rest energy, zero speed of an apparently motionless frame of reference).
- Zero establishes quantum identicality.

  Particle identicality is achieved when the difference between one particle and another is zero. Identicality is only achieved by quantum objects, and zero represents the physical measure of the difference among identical objects.
- As with all other physical constants, attempts to measure its value exactly, must fail. The zero constant is mathematically certain but has a physical uncertainty, like all things physical.

  For example, there cannot be a measured absolute zero temperature because it would require molecules to be in a state of zero energy, without movement, and we cannot measure zero energy (the Heisenberg uncertainty principle). Furthermore, it would be impossible to design a thermometer that had non-zero energy that was capable of measuring a zero energy object without imparting some of its energy to the object during the measurement process (rendering the object's temperature non-zero).

Do all of the mathematical constants qualify as fundamental physical constants? Some are derivative of other (more fundamental?) numbers and should probably be excluded from our list. Other mathematical constants have no known relationship to any physical laws or processes. Still, physicists have long marveled that mathematical relationships often dovetail with the fundamental laws of physics, in unexpected ways. For further discussion, readers may enjoy an essay written by theoretical physicist Eugene Wigner titled "The Unreasonable Effectiveness of Mathematics in the Natural Sciences." [34]

[34]Eugene Wigner's essay is published in Communications in Pure and Applied Mathematics, Volume 13, New York: John Wiley and Sons, 1960.

4.9 Proving nothingness has physical properties

"If people never did silly things, nothing intelligent would ever get done".
—Ludwig Wittgenstein

Even "nothingness" has properties. If nothingness had no properties, then it would have the property of having no properties! Aside from the aforementioned glib "proof", we can easily show that nothingness (i.e., empty space) has all sorts of properties. In fact, the symmetries that account for reality all arise from nothingness.[35]

Let's return, for a moment, to the section titled "Proving there is no ether." We recall that after the Michelson-Morley experiment discredited the notion of the ether (1886), scientists were unable to fully explain reality. How, they asked, were electromagnetic radiation and gravity carried through space if there was no ether to do the carrying? Physicists at the time had no credible way to thinking about forces (e.g., electromagnetic radiation, gravity) acting at a distance with nothing in between to deliver the effect.

Einstein was more than ready to drop the notion of the ether as the carrying medium for light, but he insisted that the universe has properties, and that the ether holds the physicality of spacetime. He insisted, in a 1920 lecture, that without the ether there would be "no possibility of existence for standards of space and time (measuring-rods and clocks), nor therefore any space-time intervals in the physical sense." [36] Einstein understood that **without the ether there would be no universal fundamental constants**, all of which specify physical qualities of the universe and are not products of forces.

To further quote Einstein,

"To deny the ether is ultimately to assume that empty space has no physical qualities whatever."
—Albert Einstein, from a 1920 lecture delivered at the University of Leiden, titled "Ether and the Theory of Relativity" (translated from the German)

Let's take a moment to review some of the many physical properties of empty space.

1. The principle of locality

   The principle of locality asserts that every quantum event, that is to say everything that happens occurs because of local conditions,

---

[35] See section titled "Proving that reality derives from symmetries".

[36] Quote from Einstein's 1920 lecture delivered at the University of Leiden, titled "Ether and the Theory of Relativity" (translated from the German).

usually forces, that cause the event to occur. Hence, space must have a physical nature that accommodates these exchanges, and we call this physicality of space "the ether."

Like all simple theories, the principle of locality has its problems. There is experimental evidence that certain types of events are non-local, happening over distances, with no lapse of time. Two examples, both closely related to one another, are the collapse of the wave function (see section titled "Proving the collapse of the wave function invalidates measurement"), discussed earlier, and quantum entanglement. Regarding entanglement, when a quantum object with spin (e.g., an electron) is created, another quantum object must be created having the opposite spin; otherwise angular momentum would not be conserved. When one of the two quantum objects commits to a particular spin value (e.g., when its spin is measured), then we can be certain that its twin must have the opposite spin. This would be true regardless of the location of the two particles. This feature is known as quantum entanglement and seems to violate the principle of locality.

If the principle of locality is violated, then can we assert that the ether must exist? Like so much of science, the confusion comes from the different ways we choose to interpret a particular phenomena. We can interpret entanglement as a purely local phenomenon (occurring entirely in the moment when paired particles are created) whose outcome is revealed at a later time. We can interpret wave collapse as a phenomenon that occurs locally, through the wave, but evaluated outside of the wave. Interpreted thusly, neither entanglement nor wave collapse violates the principle of locality. All observations are subject to interpretation.

2. Permittivity and permeability

   The speed of light in a vacuum is determined by the permittivity (ε_0) and the permeability (μ_0) of space,

$$c = \frac{1}{\sqrt{\varepsilon_0 \mu_0}}$$

   Light travels through the nothingness of space. If nothingness had no properties at all, then there would be no resistance to the propagation of light, and its speed would be without limit. Because empty space has the properties of permittivity and permeability, both of which being fundamental constants and finite, the speed of light is constant and finite, as measured in all frames of reference.

3. The physicality of particles made of nothing

   Fundamental particles are composed of nothing (i.e., there is no "stuff" of which fundamental particles are composed). This tells us that nothingness expresses the observable properties of particles.

4. Gravity

   Gravity is a deformation of space, which would not be possible if empty space lacked the property of deformability.

5. The expansion of the universe

   Hubble, in 1929, showed that the universe is expanding in all directions. The notion of expansion implies that there is something into which the universe is expanding. The deformability of empty space, and the expansion of the universe both suggest that nothingness is a type of fabric, with geometric properties.

6. Energy Fluctuations

   The Casimir effect proves that energy fluctuations must exist, even in "nothingness." [37] This implies that nothingness has observable physical properties.[38]

7. Magnetic and electric potentials

   The magnetic potential and the electric potential both persist in the absence of fields. Even when we remove fields, we have measurable potentials (electric and magnetic), both of which are basically mathematical abstractions. This would suggest that fields are not fundamental and that potentials (originally contrived as mathematical abstractions) are products of an ether.

8. Unruh effect and the relativity of nothingness.

   The aforementioned Casimir effect produces energy fluctuations in the quantum fields of empty space, and these energy fluctuations produce short-lived and unobservable particles typical of their respective fields. We refer to these particles as being "virtual" in the sense that they cannot be observed, but they can be incorporated into the calculations of particle physics (as mediators of particle interactions).[39]

[37]For a discussion of the Casimer effect, see section titled "Proving energy fluctuations must exist everywhere in space".

[38]The energy fluctuations that arise from nothingness are observed in the aforementioned Casimir effect, the Lamb shift, and symmetry breaking, the latter being responsible for quantum tunneling, spontaneous photon emissions (as occur in lasers), and predicted cosmological phenomena such as black hole evaporation).

[39]See section titled "Proving that fundamental particles do not collide with one another".

The Unruh effect (short for the Fulling-Davies-Unruh effect) is the prediction that an accelerating observer moving through empty space will observe an increasing temperature. The temperature increase, which would not be observed by an observer at relative rest, would be due to the effect of friction between vacuum energy fluctuations and the molecules of the accelerating observer.

No convincing evidence for the Unruh effect has yet been accepted. Still, its theory seems reasonable. If proven true, the Unruh effect would be one more reason to believe in the physicality of nothingness.

9. Geometry

   Geometric properties of empty space exhibit symmetries, and symmetries determine the laws of physics.

We conclude that the ether is no less a legitimate component of our universe than particles, fundamental constants, and physical laws. The long-abandoned notion of the "ether" has resurfaced as the fabric of reality.

4.10 Proving Gödel's incompleteness theorem (incomplete form)

"All the greatest and most important problems of life are fundamentally insolvable. They can never be solved, but only outgrown".
—Carl Jung

It would seem obvious that all true statements must be provable; otherwise we lose our ability to distinguish reality from unreality. Indeed, science is built on the premise that assertions of truth are testable. It may come as a shock to know that not everything that is true is provably true, and not everything that is false is provably false.

The mathematician and philosopher Kurt Gödel published in 1931 his "incompleteness theorems," proving that there are limits to what we can say about our universe. His first incompleteness theorem stated that "any consistent formal system F within which a certain amount of elementary arithmetic can be carried out is incomplete." Simply put, there are assertions that can neither be proven nor disproven. Gödel devised a clever, but somewhat convoluted way of proving his theorem that we won't try to summarize here. It suffices to say that the key step of his proof involves a self-referential statement that poses a logical contradiction. Basically, the statement "All true assertions can be proven to be true," establishes a contradiction that nullifies the proof of the statement.

Let's dawdle a bit to imagine where Gödel is leading us. Let's agree that all statements must be either true or false, and we'll begin with a statement that we shall call "Q" that asserts that Q cannot be proven.

1. Q asserts that Q cannot be proven.
2. We'll assume that Gödel was just plain wrong and that all true statements can be proven to be true.
3. If we assume that Gödel was wrong, then Q can be proven to be true whenever Q is true.
4. But the statement Q asserts that Q cannot be proven.

   Therefore,
5. When Q is true, then Q cannot be proven to be true.

   Therefore,
6. When we assume that Gödel's Incompleteness Theorem is wrong,

we run into a contradiction, thus proving that Gödel's Incompleteness Theorem is correct.

In other words, the assertion "All true assertions can be proven to be true," must be false.

In one stunning manuscript, Gödel managed to create an abstract limit on what we may know about our real universe.[40]

---

[40]Gödel K. Uber formal unentscheidbare Satze der Principia Mathematica und verwandter Systeme, I (On formally undecidable propositions of Principia Mathematica and related systems). Monatshefte fur Mathematik und Physik 38:173-198, 1931.

4.11 Proving the unprovability of randomness

"I returned, and saw under the sun,
that the race is not to the swift,
nor the battle to the strong,
neither yet bread to the wise,
nor yet riches to men of understanding,
nor yet favour to men of skill;
but time and chance happeneth to them all".
—Ecclesiastes

When we speak of randomness, we are considering a process that can be modeled as a sequence of characters that have been generated by a totally unpredictable process. A coin toss is sometimes offered as a random process. With each flip of the coin, you can see a heads or a tail. If heads, we'll say that the character generated is "1". If tails, we'll say that the next sequential character is "0". After a few million tosses, we'll have a long and seemingly random sequence of zeros and ones. At first blush, nobody can look at any one character in the sequence and guess with more than 50% accuracy the value of the next character in the sequence.

Of course, coin tosses are a crude and inadequate way to create a random sequence. Every coin toss has a preferential outcome, based on the physical attributes of the coin, how the coin is held, how the coin is flipped, and so on. A skilled gambler can toss a coin that will land on any chosen side with much greater than 50% odds.

Here's a simple inductive proof to show that, given any sequence of characters, there is no way to prove that the sequence is truly random (even if it were so).

1. Begin with some physical process that selects a "1" or a "0" as its output.[41, 42]
2. Once we have our putatively random sequence of 0s and 1s, we can pull out the first character in the sequence and ask ourselves whether it was produced by a random process.

[41] We keep in mind that any sequence of characters can be converted into a sequence of ones and zeros, so we might as well use these characters in our proof.

[42] When we refer to physical processes, we can use frequency samplings of black body radiation, time intervals between radioactive decay emissions, winning lottery numbers from the past 20 years, and so on. Seemingly random number sequences (such as the first 10,000 decimal places of pi) can be represented as character strings that can be converted to base 2 (i.e., sequences of 0 and 1).

3. We quickly see that there really is no way, just by observing the character, to determine whether the character was generated randomly. We say that we need to look at the next character to see if the first character is random.

4. We pull the next character and ask ourselves whether the first two characters were selected randomly. We still cannot draw a conclusion. We need additional characters, with which we will look for a random pattern, and from this larger set of characters, we'll make our determination.

5. We begin to add the subsequent characters of the string, in order. With each additional character, we find ourselves a bit more confident that we are looking at a random sequence of 0s and 1s, but we are never certain that we have randomness.

6. Adding more characters to the sequence doesn't really help. **Any finite stretch of 0s and 1s that seems to be random may have been generated by a non-random number generator that will eventually repeat the same sequence (in a very non-random fashion), if we wait long enough.** [43]

   Therefore,

7. We cannot prove that any finite sequence of numbers is random.

8. But, it is impossible to review an infinite sequences of numbers.

   Therefore,

9. Randomness is unprovable.

Perhaps there are mathematical (i.e., nonphysical) operations that yield random outcomes. For example, the Perl script below was used to generate a sequence of prime numbers.

```
print ‘‘2,3,";
for($i=4;$i<1000000;$i++)
  {
  for $thing (2 .. int(sqrt($i)))
    {
    $state = 1;
    if ($i \\% $thing == 0)
      {
      $state = 0;
```

[43]Pseudorandom number generators will repeat their output sequence after a large, but finite, output, and the length of the repeating output is known as the generator's period. The Mersenne twister is a popular (pseudo)random number generator (named for Marin Mersenne), having a period of $2^{19937} - 1$.

```
        last;
        }
    }
print ''$i\,'' unless ($state == 0);
}
```

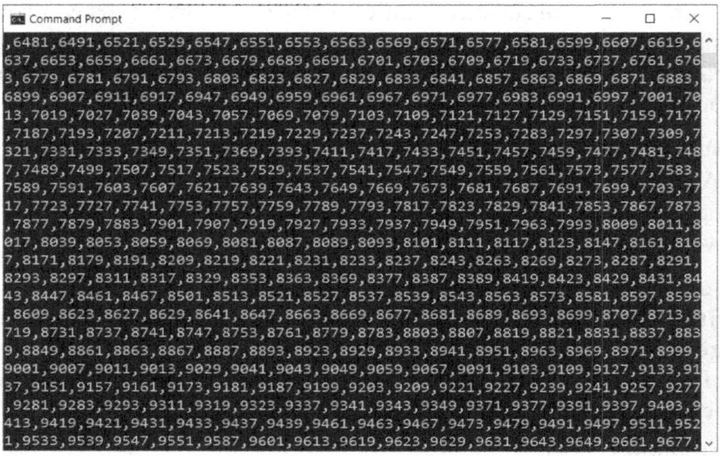

```
Command Prompt
,6481,6491,6521,6529,6547,6551,6553,6563,6569,6571,6577,6581,6599,6607,6619,6
637,6653,6659,6661,6673,6679,6689,6691,6701,6703,6709,6719,6733,6737,6761,676
3,6779,6781,6791,6793,6803,6823,6827,6829,6833,6841,6857,6863,6869,6871,6883,
6899,6907,6911,6917,6947,6949,6959,6961,6967,6971,6977,6983,6991,6997,7001,70
13,7019,7027,7039,7043,7057,7069,7079,7103,7109,7121,7127,7129,7151,7159,7177
,7187,7193,7207,7211,7213,7219,7229,7237,7243,7247,7253,7283,7297,7307,7309,7
321,7331,7333,7349,7351,7369,7393,7411,7417,7433,7451,7457,7459,7477,7481,748
7,7489,7499,7507,7517,7523,7529,7537,7541,7547,7549,7559,7561,7573,7577,7583,
7589,7591,7603,7607,7621,7639,7643,7649,7669,7673,7681,7687,7691,7699,7703,77
17,7723,7727,7741,7753,7757,7759,7789,7793,7817,7823,7829,7841,7853,7867,7873
,7877,7879,7883,7901,7907,7919,7927,7933,7937,7949,7951,7963,7993,8009,8011,8
017,8039,8053,8059,8069,8081,8087,8089,8093,8101,8111,8117,8123,8147,8161,816
7,8171,8179,8191,8209,8219,8221,8231,8233,8237,8243,8263,8269,8273,8287,8291,
8293,8297,8311,8317,8329,8353,8363,8369,8377,8387,8389,8419,8423,8429,8431,84
43,8447,8461,8467,8501,8513,8521,8527,8537,8539,8543,8563,8573,8581,8597,8599
,8609,8623,8627,8629,8641,8647,8663,8669,8677,8681,8689,8693,8699,8707,8713,8
719,8731,8737,8741,8747,8753,8761,8779,8783,8803,8807,8819,8821,8831,8837,883
9,8849,8861,8863,8867,8887,8893,8923,8929,8933,8941,8951,8963,8969,8971,8999,
9001,9007,9011,9013,9029,9041,9043,9049,9059,9067,9091,9103,9109,9127,9133,91
37,9151,9157,9161,9173,9181,9187,9199,9203,9209,9221,9227,9239,9241,9257,9277
,9281,9283,9293,9311,9319,9323,9337,9341,9343,9349,9371,9377,9391,9397,9403,9
413,9419,9421,9431,9433,9437,9439,9461,9463,9467,9473,9479,9491,9497,9511,952
1,9533,9539,9547,9551,9587,9601,9613,9619,9623,9629,9631,9643,9649,9661,9677,
```

FIGURE 4.1: A progression of prime numbers, beginning with 6481. Notice that every prime ends with 1,3,7, or 9.

The prime numbers may seem random. Not so. With the exception of two small primes (2 and 5), all prime numbers end in a 1,3,7, or 9. The "random" numbers generated from an algorithm are deterministic, meaning that they are determined by the algorithm, and it is difficult to imagine that anything can be simultaneously determined and random (indeterminate). Indeed, when we scrutinize sequences of putatively random numbers generated by mathematical algorithms, we can show that, eventually, the list of numbers will repeat itself in a most non-random fashion.[44] Having so stated, we should all be pleased to know that mathematicians have succeeded in creating strong pseudorandom number generators that serve the same purpose as truly random numbers.

If mathematical algorithms cannot produce truly random number sequences, are there physical processes that meet the requirements for true randomness? There are certainly a variety of processes that would seem to be random (as discussed in the section titled "Proving determinism, under the assumption of causality". For example:

- Radioactive emissions (i.e., neutron decay events)

[44]Green B. Tao T. The primes contain arbitrarily long arithmetic progressions. Annals of Mathematics 167:481-547, 2008.

These are weak-force nuclear decay events, such as those sensed by a Geiger counter. We know that they occur, but we cannot predict when any single event will happen.

- White noise

  Natural signals are never pure. There is always a low level of noise and the noise is characterized by a seemingly random sequence of pulses.

- Brownian motions

  Molecules bounce around in every possible direction, colliding with other molecules, with unpredictable paths.

- Black body radiation

  At any given temperature, objects emit a continuous spectrum of light waves, independent of the source, and thus carrying no information about the source of emission (i.e., random values)

- Vacuum energy/field fluctuations

  When all of the energy and matter are removed from a volume of space, we see small fluctuations of energy. Because vacuum state fluctuations occur spontaneously, without an energy source or a preceding event that accounts for their existence, they are thought to occur randomly.

The problem with supposedly random physical expressions of randomness are two-fold.

First, all we know about a supposedly random physical process are our measurements of the outcomes. Measurements are always biased by the limitations of our measurement equipment (i.e., no instrument is 100% accurate). Therefore, the act of measurement is always non-random. Therefore, **any measured outcome of a truly random process (assuming any exist in nature) must be non-random.**

Secondly, once we have the results of a supposedly random process (i.e., a long list of seemingly random numbers), we still need to verify that the list is random. If we have no way of proving that the list is random, then we really cannot be certain that we have succeeded in producing anything that is truly random. The next best thing to a test for randomness is a test for non-randomness. If we can determine, with certainty, that a sequence is not non-random, then we can infer that the sequence must be random. **Unfortunately, tests for non-randomness are based upon statistical reasoning, and do not provide us with any certain knowledge that a sequence is truly random.** [45]

[45]The U.S. National Institute of Standards and Technology has published a set of tests to determine randomness. Technically, these are statistical tests designed to determine the likelihood that a sequence is non-random. If a sequence is highly unlikely to be non-random, the inference is drawn that the sequence is likely to be random. See NIST Publication 800-22, A Statistical

What does it matter whether true randomness exists? Actually, the consequences of a universe operating under true randomness are profound.

In a universe in which randomness exists, we might see the following.

- The evolution of our universe could not determined by Schrödinger wave equation.

  Random events would interfere with predictions stemming from the wave equation.

- Determinism would not exist.

  Our past, present, and future could not have been determined at the moment when the Big Bang occurred.

- The black hole information paradox would disappear.

  The black hole information paradox asserts that information entering the black hole is lost through Hawking radiation (see section titled "Proving that information is a conserved property" for background and discussion of this fascinating topic). The black hole information paradox is only a paradox if the universe is deterministic (i.e., predictable with certainty). If there is randomness in the universe, then loss of information through Hawking radiation of black holes would not disrupt the predicted evolution of our universe (because the evolution of the universe would be unpredictable). Hence, if there is randomness in the universe, then information is not conserved and the black hole information paradox disappears.

- The law of conservation of information would be broken.

  The evolution of the universe from one moment to the next would not be wholly determined by the positions and velocities of every point in every field (randomness would intervene). Hence, information would not be conserved from one moment to the next.

- Random energy fluctuations might attain any size.

  If energy fluctuations were truly random, the magnitude of a fluctuation would be unpredictable, and could theoretically attain any size, without limit. If this were the case, energy could arise randomly at any point in space, sufficient to create a singularity, and a singularity of sufficient size could spawn the creation of a Big Bang.

- Free will may exist.

  We may actually be responsible for our own actions.

Test Suite for Random and Pseudorandom Number Generators for Cryptographic Applications, 2010.

4.12 Proving spacetime is the carrier of causality

"The difference between theory and practice is that in theory, there is no difference between theory and practice".
—Richard Moore

Causality is the notion that all events must have a cause. Based on our observations of reality, causality is local, meaning that the causal event must occur at the same time and location of its effect. For example, if a pin pops a balloon, the pin must actually touch the balloon, immediately preceding the "pop". A sequence of causal events evolves over time and space, with each preceding event occurring at the time and location of its cause. This state of affairs is sometimes referred to as the law of locality or as the rule of no action-at-a-distance. Because causality must meet certain criteria of space and time, we can infer that spacetime and causality are intimately related. In fact, it is easy to see that spacetime is the carrier of causality.

- Before we can have a spacetime event, we must first have causality.
- Before we have a causal event, we must have spacetime.
- But we do have spacetime and we do have causal events
- This is a paradox that can only be solved by unifying spacetime with causal events (see section titled "Proving the general solution to precedence paradoxes")

  Therefore,

- Spacetime mediates causal events and can be thought of as the carrier of causality

Knowing what we know about spacetime-causality, can we construct a system by which we can prove that information is a conserved quantity? We'll start by defining the meaning of a blockchain and then creating a blockchain that records all of the successive events that occur in the universe. We begin with the assertion that the universe evolves over time via events that are associated with other events, through causality. This tells us that the universe can be ordered by its events. The tool by which events can be strictly ordered is the blockchain.

Let's review what it means to be a blockchain, maintained as software. At its simplest, a blockchain is a collection of short data records, with each record consisting of some variation on the following :

```
<head>--<message>--<tail>
```

Here are the conditions that the blockchain must accommodate:

- The head (i.e., the first field) in each blockchain record consists of the tail of the preceding data record.
- The tail of each data record consists of a one-way hash of the head of the record concatenated with the record message.
- Live copies of the blockchain (i.e., a copy that grows as additional blocks are added) are maintained on multiple servers.
- A mechanism is put in place to ensure that every copy of the blockchain is equivalent to one another, and that when a blockchain record is added, it is added to every copy of the blockchain, preserving the same record sequence, and the same record content.

We won't discuss the implementation of blockchain architecture or the various uses of the blockchain paradigm (e.g., electronic ledgers, time-stamping authority, conducting impartial elections with undeniable results, and currency management). Much has been written on the topic, and interested readers can consult other sources. Here, we are interested only in the general properties of blockchains:

1. The blockchain conserves information.

   Every blockchain header contains the values in the entire succession of preceding blockchain links, holding the information of every event.

2. The blockchain is immutable.

   Changing any of the messages contained in any of the blockchain links would produce a totally different blockchain. Dropping any of the links of the blockchain or inserting any new links (anywhere other than as an attachment to the last validated link) will produce an invalid blockchain.

3. The blockchain is recomputable.

   Given an unordered list of the same head-message-tail entries found in the original blockchaing, the original blockchain, with all its headers and tails arranged in proper sequence, can be rebuilt. If it cannot recompute, then the blockchain is invalid.

4. The blockchain is a trusted ordering or events.

   Our blockchain does not tell us the absolute time that a record was created (because there is no such thing as absolute time), but it gives its order of occurrence. sandwiched as it is between preceding and succeeding causal events.

By the following reasoning, we can show that the universe can be modeled as a blockchain.

1. Action-at-a-distance tells us that an effect cannot happen without a cause.
2. But every cause is itself an event.
3. Therefore, every cause can be traced back to a prior cause and every prior cause can be traced back to its prior cause, and this process can continue to the first event of the universe.
4. Therefore, the information of the blockchain contains all the information necessary to advance the universe from event to event, and all of the information needed to rewind the universe back to its earliest moment.
5. A blockchain is a conserved informational construct (i.e., the blockchain construction guarantees that no information is lost over time).
6. Therefore, the information describing the sequence of events of the evolving universe is a conserved property.

It all seems pretty much cut and dry, but we have overlooked two very important details.

- We have not proven that space and time are calibrated across the universe.
- We have not proven that all events occur through causality.

Let's review a gedanken experiment that violates spacetime calibration and causality, thus undermining our proposed universal blockchain.[46] In preparation for our experiment, a little background is necessary. Fundamental particles are created in pairs, and the quantum state of the paired particles includes spin. In order to conserve angular momentum, the measured spins of the two particles must be opposites (e.g., if one particle has a measured spin "up", then the other particle of the pair must must have spin down). Prior to measurement, both particles exist in a ambiguous quantum states (i.e., both particles exist in states up and down).[47] When the spin of either particle is measured, its quantum state is "fixed" (either up or down, but not both). At

[46]The experiment was first described in 1909 by Albert Einstein, Boris Podolsky, and Nathan Rosen is known as the EPR paradox. The original manuscript was published as: Einstein A, Podolsky B, Rosen N. Can quantum-mechanical description of physical reality be considered complete? Phys Rev 47:777-780, 1935.

[47]This is the Copenhagen Interpretation of the Schrödinger wave equation, a collection of views by leading quantum physicists (dominated by Niels Bohr) that first gained traction in 1920.

the moment that one particle's spin is measured and "fixed", the other particle must take the opposite spin (to conserve the angular momentum of the universe).

In our experiment, the paired particles may be located at a vast distance from one another. When we measure the spin of one of the paired particles, the spin of the other particle is instantly committed to the opposite value. Because the causal event occurs over a distance, spatial causality is violated.

Let's extend the experiment by putting one of the two particles in a spaceship traveling at near-light speed, while its paired particle is relaxing on planet Earth. The clocks on Earth are moving at a different speed from the clocks on the spaceship. Imagine that we perform our experiment on January 1, 2080 by Earth time, which happens to be Jan 1, 2070, by spaceship time. On Earth (in the year 2080), we measure the spin of one of the paired particles, and we find that is in the "up" state. Simultaneously, on the spaceship (in the year 2070), the paired particle transitions from an ambiguous (up/down state) to a fixed "down" state. Therefore, an event on Earth produces an effect on the spaceship ten years before the event happens (according to their respective clocks). This is an apparent violation of temporal causality.

How would a universal blockchain deal with the discrepancy?

As we previously noted, a blockchain is a trusted **ordering** or events that does not tell us the time or place that a record was created. In the case of the gedanken experiment, the time discrepancy (between Earth and spaceship) would not change the event's blockchain entry. There would be one entry at the moment when the spin values of the paired particles spins were determined. Another way of expressing the situation is that space and time (i.e., rulers and clocks) carry causality but cannot be calibrated. Blockchains carry the sequence of causality but do not require calibration.

4.13 Proving the collapse of the wave function invalidates measurement

"It is difficult to find a black cat in a dark room,
especially if there is no cat"
—Chinese proverb

Measuring the spacetime location of a wave is no easy task. Attempting to do so draws us into the so-called collapse of the Schrödinger wave equation.[48] In brief, we have the following situation. When we pass light through a slit, a wave diffraction appears on the screen behind the slit, making visible periodic wave peaks and wave troughs. The reason we see these peaks and troughs is that the edges of the slit cause the light (at those edges) to bend and spread out, producing multiple spreading wave patterns that, when combined on the screen, will reinforce or cancel periodically.

When we pass light through **two** slits, we see a somewhat different diffraction pattern, often described as a wave interference pattern. The interference pattern is formed when the diffracted waves exiting the two slits intersect, and the peaks and troughs of the two diffracted waves cancel out, producing a repeating pattern on a screen positioned behind the slits.

When we pass an electron beam through two slits, we see the same pattern as we would find with a beam of light, or a water wave passing through two slits and colliding on the other side, an indication that particles have the properties of waves.

If we were to modify the experiment slightly, by placing a light sensor in the path of the wave as it approaches one of the slits, then the light passing through the slits no longer behaves like a wave. Instead, it behaves like a beam of photon particles. The screen behind the two slits measures exactly two columns of impacted particles (i.e., the repeating interference pattern is gone). The placement of a detector between the light source and the slits was the only condition that changed in the two experiments. Apparently, the act of measurement instantly transformed the waves into particle beams! [49]

[48] We'll be discussing the Schrödinger wave function in section titled "Proving Schrödinger's wave equation".

[49] The instantaneous transformation occurs at both slits, including the slit at which no detector was placed! How did the particles at the second slit know that the particles at the first slit were being sampled and measured? Was this an example of action-at-a-distance (i.e., an effect produced without a carrier). If so, was this action-at-a-distance instantaneous? Einstein referred to this phenomenon as "spooky action-at-a-distance." A possible explanation is that the waves moving through the slits are interacting (via interference) and are part of the same system. The wave collapse occurs in the wave system, not in individual photons at separate locations.

In the quantum world, the behavior of particles, from one moment in time to the next, is determined by the Schrödinger wave function. The wave function models particles as harmonic oscillators, and the position of the particle at any given moment falls anywhere within a range of probable locations. The particle, modeled as a wave, is said to occupy all allowable quantum states simultaneously. When the wave is measured, the Schrödinger wave function ceases to hold, and the function is said to **collapse** into one specific quantum state (i.e., the value of the measurement). Hence, the act of measurement changes what we are attempting to measure (by converting it from a wave probability to one particular value).

Summarizing,

1. The wave equation embodies the evolution of the system (i.e., the system consisting of waves in fields).
2. The act of measurement (i.e., placing a light sensor in the path of the wave) collapses the wave equation.
3. Because measurements collapse the wave equation, we conclude that measurement alters the system being measured.
4. If the system being measured is altered by measurement, then measurements of the system are invalid.

   In other words, valid measurements are impossible to achieve.

This is the so-called measurement problem of quantum physics.

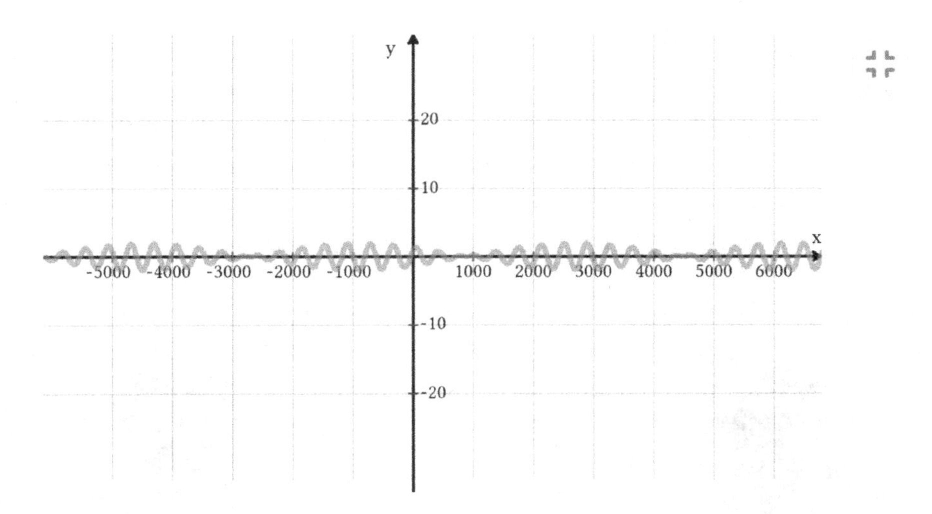

FIGURE 4.2: When waves are added together, as happens when two waves meet as they exit from a double-slit, the peaks and valleys of each respective wave will interfere with one another, producing a composite wave with areas of low amplitude alternating with areas of increased amplitude. As an example, our figure shows the resulting interference wave pattern that results when a sine wave intersects a cosine wave ($cos(x) + sin(0.9x)$).

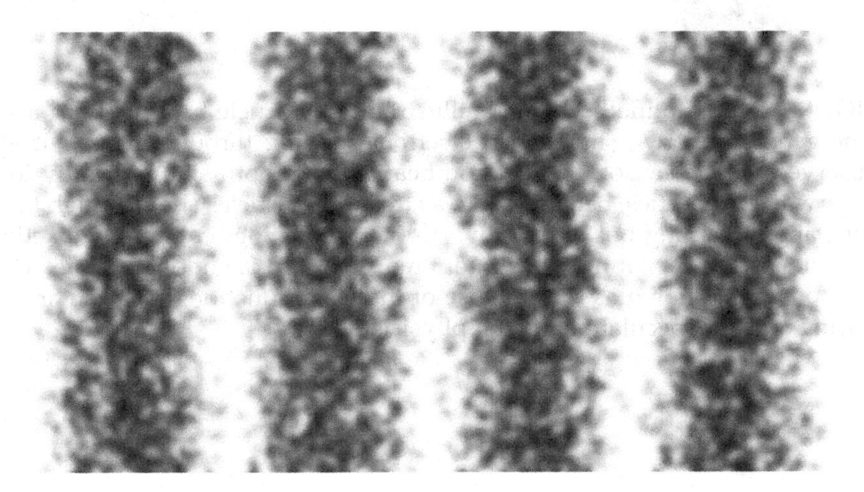

FIGURE 4.3: Simplified schematic illustrating a wave interference pattern after emergence from two slits. When we send a beam of electrons through two slits. After the waves exit the slits, we see an interference pattern characterized by repeating dark areas (wave reinforcement) separated by light areas (where the waves cancel out) displayed on a screen. This finding indicates that electrons have wave properties.

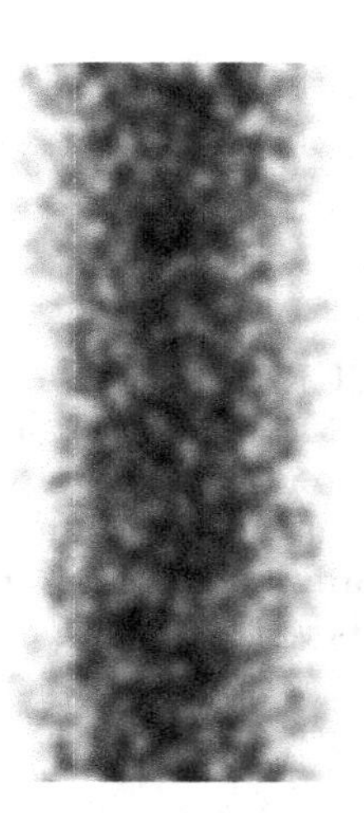

FIGURE 4.4: Simplified schematic illustrating a particle beam pattern after emergence from two slits. When electrons moving through either slit are observed with a sensor, the electron beams (from both slits) instantly lose their wave properties, behaving like a stream of particles shot through both slits, and streaming out in exactly two (non-repeating) columns impacting the screen. The repeating wave interference pattern is suddenly gone! This phenomenon, known as the collapse of the wave function, has been an enduring topic of speculation among physicists.

4.14 Proving fundamental constants could not have arisen from a causal event

"The best that most of us can hope to achieve in physics is simply to misunderstand at a deeper level".
—Wolfgang Pauli

The fundamental constants of nature dictate the physical nature of the universe. This being the case, we cannot help but wonder how these constants came into existence. Surprisingly, there is a simple proof that the fundamental constants have no cause.

1. The fundamental constants of nature are dimensionless (see section titled "Proving that all fundamental constants are dimensionless").
2. Because the fundamental constants are dimensionless, they are not associated with any particular time or place, and are timeless (existing at all times) and ubiquitous (existing throughout all of space).
3. Causality requires that a causal event must occur at the same time and at the same location as the resulting effect of the cause.
4. Since constants have no assignable time or place, they cannot have resulted from a causal event.

If the fundamental constants were not "caused" by anything, then how did they come to be? Presumably, the constants arose as conditions of symmetries. The symmetries, in turn, set the conditions upon which reality emerges (see section titled "Proving that the operating system of the universe is determined by symmetries".)

4.15 Proving determinism, under the assumption of causality

"To arrive at abstraction, it is always necessary to begin with a concrete reality . . . You must always start with something. Afterward you can remove all traces of reality".
—Pablo Picasso[50]

In a pre-determined universe, we may think we have free will, but our every action was determined at the moment that the universe began. Here is the argument for determinism:

1. In a causally dependent universe, all events occur as the consequence of some preceding event. That is to say, nothing happens without a cause.
2. If this were the case, then every causal event must itself occur as the consequence of some preceding event.
3. If so, every event that occurs at the present can be traced backward, in a causality chain, to the very first event that has ever happened.
4. If that is the case, then every event that has ever occurred, and every event that will occur in the future, was predetermined by the first event that occurred in our universe.

A second argument for determinism arises from a popular interpretation of the Schrödinger wave equation.

1. The Schrödinger wave equation applies to a linear system.
2. In a linear system, the components are additive.
3. Everything in the universe is composed of waves, and every wave has its own Schrödinger wave equation that determines its evolution from a time period integrated over $-\infty$ to ∞.
4. The evolution of the universe is determined by adding all of the Schrödinger wave equations (a property of linear systems).
5. Therefore the evolution of the universe over time is pre-determined by its wave equation.

[50]Quoted in Robert S. Root-Bernstein and Michele M. Root-Bernstein. Sparks of Genius. Mariner Books, 1st edition, 2001.

The arguments against determinism are based upon causal agnosticism (i.e., we do not know whether causality applies to every event in the universe), to wit:

1. The very first event of the universe (i.e., the cause of everything that follows) has no cause. If the first event had a cause, then it would not qualify as the first event. We are asked to accept a first causal event for the universe on faith alone, but that's not what scientists do. If the first event of the universe had no cause, then we can't really say that the universe is deterministic, because the first event, having no cause, would violate determinism.
2. If there are events that occur randomly, then those events could serve as causes for subsequent events. Random events are, by definition, unpredictable and indeterminate. Therefore, randomness (if it exists) violates determinism.

   Although randomness can never be proven (vida infra), there are various physical events whose timing or ordering cannot be predicted. So far as anyone can tell, they seem to be random (i.e., not pre-determined).

   These include:

   - Neutron decay events
   - Black body radiation
   - Thermal noise
   - Brownian motion
   - Johnson noise (due to thermal agitation of charge carriers)

Finally, regarding the argument for determinism based on the application of the Schrödinger wave equation to the universe as a whole, we might offer that the Schrödinger wave equation is, at best, an intelligent guess. It seems to give a fair account of many observations of the quantum world, but it leaves us with unanswered questions (such as the collapse of the wave equation). We need not take the Schrödinger wave equation too seriously.

In summary, It would seem that if every event in our universe has been preceded by a cause (including the root event), then the universe is deterministic. Otherwise, we live in a universe where **things just happen**.

4.16 Proving information is a conserved property

"The ability to reduce everything to simple fundamental laws does not imply the ability to start from those laws and reconstruct the universe".
—PW Anderson

The conservation of information is certainly one of the more metaphysical concepts in physics. Explanations often sound mystical and unconvincing. Let's begin with a grand vision of the law, from which we can drill down to a quantum example. The Schrödinger wave equation, in principle, describes a purely deterministic universe in which existence from moment to moment, and the information describing each moment, is fully described. Hence, every moment in the evolution of the universe holds all of the information necessary to advance the universe to the next moment in time. Put very simply, the Schrödinger wave equation enforces the conservation of information.[51] Of course, we do not have any real proof that the Schrödinger wave equation is true and inviolate. So we really cannot use the Schrödinger wave equation to prove that information is conserved.

Perhaps we can find something more substantive with which we can make our point. Imagine two causal events, A and B, that happen to produce the same effect, which we'll call C. When we inspect C, we cannot always determine whether C occurred as the result of A or whether it resulted from B. For example, if I wake in the morning to find an empty coffee cup on my night table, I have no way of knowing, just by looking at the cup, whether I had left it there the prior evening, or someone else has placed the cup on the night table during the night. Either cause may have produced the same event. Therefore, outcomes are not always reversible (because several different causes may produce the same outcome). In such cases, the information regarding the sequence of events leading to the outcome had been "lost" to the universe (i.e., not conserved).

But we know that quantum events can occur forwards or backwards. We also know that non-quantum events are just the complex aggregate of quantum events. Therefore, all causal events should be reversible. Therefore, the information that describes the state of the universe prior to the occurrence of C must exist, so that we can properly reverse C and determine unambiguously which event (A or B) led to C's causation. Put in other words, the universe cannot follow the wave function's deterministic course (i.e., a course that can

[51]When physicists discuss information, they are generally considering the information that describes quantum states. However, the universe evolves from quantum states, and the information that describes quantum states can be added together to describe everything in the universe.

be traced forwards and backward) unless the totality of information describing the state of the universe is conserved from moment to moment.

Black holes seem to violate the conservation of information. As an object crosses a black hole's event horizon, it exits the observable universe. As the physical object disappears into the black hole, so too does all of the information associated with that object (i.e., the information that is necessary to describe what happens to the object as it moves from one moment of time to the next). This information that disappears into the black hole is not necessarily lost. Perhaps it exists in a state of limbo, within which the law of conservation of information is held in suspense, without being violated. Stephen Hawking has predicted that over time, black holes slowly release black body radiation. Eventually, over many eons, the leaking radiation leads to the eradication of the black hole.[52] Black body radiation is, so far as we can tell, totally random (i.e., incapable of carrying information). Therefore, we can expect that as the black hole has decayed, its contained information is slowly lost. In doing so, the conservation of information is violated.

In the case of the black hole erasure of information, physicists have a nifty theorem up their sleeves: the "No-Hiding theorem". This theorem states that when information is expunged from a system (as would be the case when black holes evaporate), the destroyed information must be distributed to some other subsystem.[53] There is some experimental evidence in support of the No-Hiding theorem. In 2011, an experiment was performed in which a nuclear magnetic resonance machine was employed to randomize the information in a qubit (i.e., the quantum state information held in the qubit was destroyed). It was found that the information lost to the qubit could be fully recovered from other qubits in the same system (i.e., the original information could not be hidden).[54]

The proof of the law of conservation of information assumes two conditions: 1) that the universe follows a predetermined course, as determined by the Schrödinger wave equation; and 2) that randomness does not exist. Of course, we do not know that the universe is deterministic, and we have no way of proving whether randomness exists. Therefore, the conservation of information does not as yet carry the same stature as a proven law of nature.

---

[52]The time required for a typical black hole to decay into nothingness, via the Hawking process, is thought to be on the order of 10^{64} years. Compare this with the age of the universe since the Big Bang, somewhat more than 10^{10} years.

[53]Put another way, you cannot hide information. When you try to hide it in one system, it will pop up in some form, perhaps as an extracted correlation, in some other system.

[54]Scholars among us can check the original paper: Samal, JR, Pati AK, Kumar A. Experimental Test of the Quantum No-Hiding Theorem. Physical Review Letters 106: 080401, February 22, 2011.

4.17 Proving fields exist

"Reality is nothing but a collective hunch".
—Lily Tomlin

The proof, such as it is, comes from our recognition of symmetries.

1. We know that symmetries exist.

   We see symmetries all around us. We can move an object from the left to right, and the object stays the same (translational symmetry). We can rotate objects without changing them (rotational symmetry). We can perform an identical experiment on Tuesday and Wednesday and expect the same outcomes (Time-shift symmetry). In addition, we have an assortment of internal symmetries that are independent of spacetime coordinates that account for electric charge, and particle spin, and the Standard Model of quantum physics.

2. We define a field as the geometry in which a symmetry operates.

3. Since the field is a mathematical construct based upon a physical reality, it would be impossible to deny its existence.

To say that fields do not exist would be much like saying that arithmetic does not exist. Basically, arithmetic (and fields) exist because our experience of reality depends upon their existence. If you find this proof unsatisfactory, I urge patience. As we learn more about symmetries and their fields, we will gain a bit more confidence in the logic of the argument.

Physicists have a cute way of summing the situation: Every symmetry has a field and every field has its own distinct quantum particle representing the smallest observable manifestation of field energy.

4.18 Proving time exists

The notion that space and time are inseparable and that either one can transform into the another, depending upon the frame of reference in which they are observed, is the cornerstone of special relativity. Just like all properties of our universe, spacetime arises as a consequence of symmetry. The symmetry transformation responsible for the existence of spacetime is "cause and effect". Every effect must have a cause, and every cause (basically every event) must have an effect, and cause-effect is reversible. When we claim that cause-effect is reversible, we mean that any cause-effect sequence could occur forwards or backward. For example, a cue ball can knock a racked set of billiard balls sending them all off in different directions. Or a set of billiard balls can collide at once, sending the cue ball back to its cue stick. The second scenario is unlikely, but we won't quibble.

There is one more caveat to the symmetry: cause-effect pairs are local, meaning that a cause occurring at one point in spacetime can exert its effect only at its immediate location (i.e., **there can be no action-at-a-distance**). Just for a moment, let's imagine that this were **not** true, as we might see in the following hypothetical sequence:

1. Let's pretend, for the moment, that we live in a universe where action-at-a-distance was possible.
2. If such were the case, an event occurring at spacetime point A could directly cause an effect at spacetime point B.

   That is to say, an effect can be **directly** caused by an event occurring at some distant place and time).
3. But we live in a universe with homogeneous spacetime, meaning that any point in spacetime is equivalent under the same set of conditions.

   This means that repeated experiments will have the same outcome here and elsewhere; now and later. Therefore,
4. If there were action-at-a-distance, the effect of a cause could occur **anywhere** in spacetime.

   That is to say that an effect can be directly caused by an event occurring at some distant place and time.

   Therefore,
5. The universe would be chaotic, unpredictable, and inhomogeneous.

Events would just **happen** without any discernible cause. For example, a planet could explode due to some action taken in the past, or the future, occurring in a distant galaxy.

6. But our universe is not chaotic, unpredictable, and inhomogeneous.

   Therefore,

7. Action-at-a-distance does not occur.

If action-at-a-distance cannot occur, then we must ask ourselves what property of the universe protects against action-at-a-distance. The answer seems to be "time." If there were no passage of time, then a causal event A could traverse the distance between A and any other point B in zero time (because time would not exist). This means that, in the absence of the passage of time, event A could produce an effect at some distant point B, thus achieving action-at-a-distance. Because an event A can never produce a direct effect at a distant point B, we can infer that time must exist. You may have heard the oft-repeated remark that "Time is what keeps everything from happening at once." This glib definition of time can be restated as "Time is what prohibits action-at-a-distance."

Having just argued that time must exist, we can ask ourselves whether there might be a set of conditions under which time does not exist. As it happens, we can reasonably assert that time does not exist for electromagnetic radiation (including visible light) and for all massless particles. Here is the logical argument that supports our assertion:

1. As discussed previously, the speed of light is an invariant in all reference frames, and all massless particles must travel at the speed of light.

2. When traveling at the speed of light, time stops.

   As a particle moves faster and faster, its clock slows down (as per special relativity). At the speed of light, time stops completely and the particle exists in the so-called "eternal now".

3. If time has stopped, then time does not exist (for particles traveling at the speed of light).

There is actually some observational data that supports this conclusion. Let's see:

- If time does not exist for massless particles, then none of the properties of the particle would change over what we, in our slower-than-light reference frame, observe as the passage of time (because time does not exist for the particle).

- If this were the case, then degeneracy could not occur in massless particles.
- In that case, we would never observe decay in massless particles (that move at the speed of light).
- Observationally, massless particles are eternal, when left alone.

  It is only when their existence is interfered with (e.g., light blocked by a mirror, high energy interactions occurring within a star) that they enter or leave their existence. Our best available estimates of the expected life-span of most massless particles is many times the known age of the universe, verging toward the infinite.
- Because **most** of the universe consists of massless objects that travel at the speed of light, the universe does not change very much, over time.[55]

  Atoms, composite particles traveling at less than the speed of light, represent a nearly negligible portion of our universe. Most of observable reality consists of massless waves traveling at the speed of light.[56] If time does not exist for massless entities, then we would expect that the universe (composed largely of massless items) to have changed very little since the Big Bang. This is exactly what we find. Today, we see the very same cosmic background radiation that was released into the transparent void approximately 180,000 years after the Big Bang. With the exception of a relatively small number of fermionic activities (e.g., star and galaxy formation, the birth of Albert Einstein, the invention of chocolate cake), nothing much has happened in the intervening billions of years.

There is a hypothetical situation for which time does not exist for objects having mass; this is the case when everything in the universe is motionless (relative to everything else in the universe). Here's the argument:

1. Observable events require movement (i.e., something that changes its distance over an interval of time).

   If nothing moves, then there is no event.
2. If the universe were composed of entities that are motionless (i.e., with no change in their distances from one another over time), then there would be no observables.

---

[55]We humans, being composed of matter, tend to exaggerate the presence of matter in the universe. In point of fact, most of the universe is devoid of matter. Electromagnetic radiation, in the form of the cosmic microwave background, dominates the void. Molecules of "solid" matter are composed primarily of ghostly gluons and quarks, the former being massless, and the latter having negligible mass. When you think about it, we barely exist at all, and the little bit of existence we have is frightfully short-lived.

[56]Not to mention dark energy and dark matter, two hypothetical sources of unobservable energy and mass.

3. If there are no observables, then there would be no observable passage of time.

   There would be no way of determining that time has passed when nothing ever changes.

4. If there were no observable passage of time, then time would not exist.

   Drawing from the definition of velocity, we could have arrived at the same conclusion. Velocity is distance traveled divided by the change in time. In a universe without motion, the distance traveled is zero and the velocity is zero, so the change in time must also be zero. Therefore, time does not exist in a motionless universe.

The notion of a motionless or frozen universe may seem far-fetched insofar as the spontaneous energy fluctuations of the universe would cause fermionic particles to move around, generating heat. Hence, there would always be some movement, and with the movement of fermions, there would be observables and the passage of time. But suppose the universe was an empty void. In this case, there would be nothing to observe, and time would not exist. We can imagine an empty and timeless universe that exists until a spontaneous energy fluctuation is sufficiently large to begin a new universe, at which point "time" makes its grand entrance.

4.19 Proving observers in different frames of reference inhabit different universes

"In astronomical terms, early in the universe translates to very far away".
—Aaron Dubrow

One of the strangest consequences of the observed expansion of the universe is that matter is disappearing.[57] Everywhere we look, galaxies are moving away from us, and the further away a galaxy is from Earth, the faster the speed at which the galaxy is receding. The same observation would be made from any location in space; the expanding universe pulls galaxies away from any observer's position. The experimental evidence for an everywhere-expanding universe was found by observing the spectral shifts of light emitted by galaxies near and far from Earth. All galaxies exhibit a Doppler shift of spectrum toward red (never toward blue) indicating that all galaxies are moving away from Earth. The more distant the galaxy, the greater the red shift, indicating a greater speed.

Although galaxies cannot move through space at speeds greater than the speed of light, there is no limit to the speed at which the universe expands. As the universe enlarges, the space between any two objects in the universe increases. Nothing "moves" in the process, but space, and the distances between objects, enlarges.

At the furthest distances from Earth, matter may recede at speeds greater than the speed of light. Anything moving greater than the speed of light becomes unobservable to us here on Earth, because the light emitted by such objects will never catch up with the speed at which it is receding from our view. As the universe continues to expand, increasing numbers of galaxies will disappear from our observable galaxy. In a fascinating manuscript, Lawrence Krauss and his colleagues, suggest that all but our most proximate galaxies will depart our observable universe in about 100 billion years.[58] Eventually, our galaxy will be entirely alone in the universe.

As it happens, the consequences of a continuous Hubble expansion are more profound than simply missing out on the beauty of a starry sky. The Hubble expansion challenges our understanding of causality and requires us to

[57]The discovery of the expansion of the universe is generally credited to Edwin Hubble (1889-1953), who published his findings in 1929. First honors are sometimes assigned to Georges Lemaître (1894-1966), who proposed a similar findings in 1927. It is not unusual for discoveries to be made independently by several scientists, at about the same time.

[58]See Krauss LM, Scherrer RJ. The return of a static universe and the end of cosmology. arXiv:0704.0221v3, 27 Jun 2007.

develop a new understanding of how the structure and content of the universe changes from observer to observer. Let's begin.[59]

All things should have a precise definition. Therefore the universe, being all things, certainly deserves a precise definition. We'll define a universe as an observable spacetime in which causality prevails. When we say that something is observable, we are saying that it produces a measurable effect. When we see something with our eyes, the light coming from an object produces a effect on the rods and cones of our eyes, which is translated by our brains into an image. Particles that are too small to visualize (such as short-lived virtual particles), may have an observable effect on particle-particle interactions, and are therefore indirectly observable. The requirement that the universe must have causality is equivalent to insisting that action-at-a-distance is impossible. A direct cause must have the same spacetime location as its immediate effect. For example, we cannot see a star unless light originating from the star reaches our eyes.

By definition, every object in our universe is directly or indirectly observable (unobservable objects are omitted from our definition of the universe). It may take a while for the light coming from an object to reach our eyes, but it will eventually arrive, given enough time. Likewise, if we have a sensitive particle detector, located somewhere in the observable universe, that is continuously broadcasting its results via radio waves, its information will come home, someday. Objects that do not emit light (such as black holes) are indirectly observable due to their observable effects on spacetime and matter in their proximity.

Now let's consider what happens with objects receding from us at speeds greater than the speed of light. For brevity we'll say that these objects reside outside our universe's horizon.

1. Objects outside the universe's horizon are invisible to us.
2. Objects that cannot be observed in our universe are no longer in our universe.[60]
3. Because such objects were, in the past, bona fide members of our universe, and because the process of universe enlargement does not alter matter (other than expanding the distance between molecules), we can infer that objects beyond the horizon do exist.

[59]What exactly happens when the universe expands? Is spacetime simply added to to the existing universe. If so, would this not prove that space and time are not conserved quantities? How does the expansion of the universe affect gravity, which is itself a manifestation of the curvature of space? These are questions that we will not be addressing here, but readers should keep in mind that the expansion of the universe has broad significance.

[60]It is ironic that in the early sea explorers feared that they might fall off the edge of the Earth, if they ventured too far. Today, astronomers speculate that objects can literally fall out of our universe, if they venture beyond the horizon of our universe.

4. If such objects exist, but do not exist in our universe, they must exist in some other universe.

5. From Earth's perspective, the universe that lies outside our horizon is expanding at a rate faster than anything in our universe.

   We can infer this because our observable universe only contains objects that are not receding from us at greater than light speeds. Hence, the objects that recede into another universe are all residing in a universe that is expanding at greater than light speeds.

6. Objects inside our universe's horizon but distant from Earth (such as an object just inside the universe's horizon, as seen from Earth), observe a different horizon than we do, and stars excluded from the universe (from Earth's perspective) is included within the universe observed from elsewhere.

   We must remember that every object in the universe sees the expansion of the universe (and the distribution of redshifts) from its own perspective, with close objects moving slowly away from their position, and distant objects moving away with great speed. From the perspective of an object just inside Earth's horizon, objects just outside Earth's horizon are fully visible and are moving slowly.

   Therefore,

7. All objects outside Earth have a horizon that is different from Earth's horizon and that includes stars and galaxies that are unobservable from Earth.

   Therefore,

8. As a result of the accelerating expansion of the universe, we can conclude that we dwell within a universe that is different from the universe of every other object.

Hence, a universe, much like everything else, is a relative reality that is understood within a particular frame of reference.

4.20 Proving that causation is not always observable

"Philosophy of science is philosophy enough".
—Willard Van Orman Quine

If we say that something "caused" the universe to come into existence, then we are saying that causation existed fundamentally, at the birth of reality. This would imply that consequences of causation are emergent properties (i.e., not truly fundamental). We humans are examples of the emergent effects of causation. It seems unlikely that we will ever have the opportunity to observe what "caused" the universe. Aside from that, can we say that all of the events that occur in our universe have an observable cause?

If we consider causality to be a fundamental feature of reality, then spacetime can be considered an emergent (i.e., non-fundamental) consequence of causality. What does this tell us about how we observe causality? Consider the following:

1. We know that the universe, currently, is expanding.

   The rate of expansion, as viewed from Earth, increases with distance from Earth (e.g., distant galaxies are receding from us at a faster speed than neighboring galaxies).

2. There is no theoretical limit to the rate of expansion of the universe, and the rate of expansion can exceed the speed of light.

   As the universe expands, the distance between objects expands, but the objects themselves do not move. Hence, an expanding universe does not alter the speed of motion of the objects within the universe and does not violate the conditions of special relativity.

3. Objects occupying a location in space that recedes from us at a rate faster than the speed of light becomes unobservable.

   This is because the light emanating from such objects will travel at a speed less than the speed with which their location in spacetime is receding from us. Thus, their emitted light will never reach us, and such objects are therefore unobservable.

4. In this case, though the expansion of the universe increases the size of spacetime, the contents of the observable universe would diminish.

   For example, galaxies may cease to exist as they recede from us at velocities that exceed the speed of light. The individual stars in distant galaxies may extinguish one after another until the galaxy disappears entirely, or the galaxy may vanish in a puff, if it is sufficiently small and fast. [61]

5. Events occurring in an unobservable galaxy (unobservable to us) may influence events in the observable universe.

   Howso?

   (a) Consider three galaxies: A, B, and C.

   (b) Suppose that galaxy C is receding from galaxy A at greater than light speed, while galaxy C is receding from galaxy B at something less than light speed, and galaxy B is receding from galaxy A at less than the speed of light.

   (c) In this case, galaxy A can observe events occurring in the galaxy B but not galaxy C. Galaxy C can observe events occurring in galaxy B but not galaxy A. Galaxy B can observe events occurring in both galaxies A and C.

   (d) Consequently, an event in galaxy C can produce an effect in galaxy B, and the effect produced in galaxy B can have an effect in galaxy A.

       The event may be as minor as an observation because a simple observation is itself an effect.

       Therefore,

   (e) An event in galaxy C (which is unobservable to galaxy A), can have an effect on galaxy A.

   Hence,

6. Causal events may not always be observable.

[61]The topic was covered in depth in the section titled "Proving observers in different frames of reference inhabit different universes".

4.21 Proving travel into the past is not possible

If time travel were possible, then we would expect that in the future, everyone will have access to their own time machines, and we would expect to see time travelers popping up in our living rooms shopping malls and restaurants. We haven't seen such visitors from the future. In fact, there is no evidence that travelers in the future have visited any sites in Earth's past. Hence, time travel is impossible.[62] Of course, this argument makes two assumptions: 1) That time travelers would have chosen to visit Earth's past; and 2) That our visitors would make themselves known to us.

A gedanken argument against time travel has been proposed by Cristopher S. Baird.[63,64]. Suppose we have a time portal that takes objects a modest three seconds into the past. We mount the portal a short distance above a trampoline, activate it, and immediately drop a ball. The ball appears three seconds before the portal is activated, adjacent to the ball that we prepare to drop. Now both balls are about to drop, but two more balls appear, also three seconds earlier. Now four balls appear just before we are about to make our drop. The balls proliferate. In this manner, an infinite number of balls appear. In no time at all, the universe is filled with balls.

If it is impossible to travel into the past, then how can we explain that events occurring in spacetime are reversible. For example, a movie taken of two billiard balls colliding can be played forwards or backwards. The collision, and its aftermath, is symmetric in time and space, meaning that the movie can be played forwards or backwards. Either versions of the movie are physically possible. Physicists have enshrined this concept of time reversibility by asserting that positrons are electrons moving backwards in time.[65]. Let's not get carried away. When we move in spacetime, we can move in any direction, include the opposite direction from which we came. When we do so, we are not moving into a realm of space in which time moves backward, any more than we are moving into a special realm of space in which distances are negative. We are simply moving to another location in spacetime. When we reverse our movie of the colliding billiard balls, we observe the movie in the forward-moving spacetime. The events move backwards in time, but the two

[62] A similar argument, known as Fermi's paradox, applies to the existence of intelligent life on other planets. If aliens were zooming around the galaxy, wouldn't they have made themselves known to us? Enrico Fermi summarized the argument in one short question: "Where are they?"

[63] Gedanken is the German word for "thought." A gedanken experiment is one in which the scientist imagines a situation and its outcome, without resorting to any physical construction of a scientific trial. Albert Einstein, a consummate theoretician, was fond of inventing imaginary scenarios, and his use of the term "gedanken trials" has done much to popularize the concept.

[64] Dr. Baird is a physicist and an active blogger. His excellent pearls of wisdom can be enjoyed at: *https : / /wtamu.edu / ∼cbaird /*

[65] This idea was first proposed by Richard Feynman

hours that we've wasted watching the movie in reverse is no different from any other 2-hour stretch of time. Similarly, positrons and other anti-particles move about in ordinary spacetime; they cannot be observed traveling into our own past.

4.22 Proving the operating system of the universe is determined by symmetries

Modern physicists search for symmetry groups from which they can infer physical reality. There is no solid proof that symmetries account for all of our reality, but for that portion of reality that we seem to grasp, there is a collection of steps that summarize how symmetries determine physical law.

1. If there is symmetry, then there is a conservation (and a law expressing the conservation).[66]

   For example, rotational symmetry accounts for the law of conservation of angular momentum. Why so?

   (a) If we know that nothing changes when an object is rotated (i.e., if there is rotational symmetry),

   then,

   (b) The momentum of the rotated object in invariant under the symmetry transformation,

   (c) If the momentum does not change, then it is conserved.

   Thus,

   (d) In a system where rotational symmetry holds, angular momentum must be conserved.

   By the same logic, it can be shown that symmetry under translation accounts for the law of conservation of momentum. Time-shift symmetry accounts for the law of conservation of energy.

2. If the quantity conserved by the symmetry is quantized, then there is a particle associated with the symmetry.

   As discussed previously, a particle is the quantum of the field.

3. If there is a symmetry, then there is a symmetry group.

   For some background on group theory and symmetries, see section titled "Proving the uniqueness of inverse elements of groups".

4. If there is a symmetry group, then there is a group isomorphism.

   A group isomorphism is a function between two groups that sets up a one-to-one correspondence between the elements of the

[66] This statement is roughly equivalent to Noether's theorem, proven in section titled "Proving Noether's theorem".

groups in a way that preserves that mapping after group operations. Because two isomorphic groups have a one-to-one mapping between their elements, and because operations on one group are preserved in the second group, then the groups have the same properties and are equivalent to one another. The rules of a group apply to its isomorphism.

Let's look at an example. We can show that the group of real numbers are isomorphic to the group of logarithms of the real numbers. There is a one-to-one correspondence among the group elements, and the operation of multiplication in the first group is equivalent to the operation of addition in the second group. When we multiply any two real numbers in the first group, we get an element that is equivalent to adding the logarithms of those real numbers (in the isomorphic group) and mapping the result back to the first group. Because logarithms are easily added or subtracted (and real numbers are difficult to multiply and divide), we might choose to conduct operations in the isomorphic group (of logarithms), and mapping the result back to the group of real numbers, when the operation is finished.[67]

5. If there is an isomorphism, then there is an irreducible representation.

   In group theory, a representation is a mapping from a group element to a vector space in such a way that the group properties are preserved. Representations are typically used to map a group to a matrix (that represents vectors). Because the group and its representations are homomorphic (i.e., being of the same algebraic type and preserving operations) scientists can directly apply the properties of the group to its matrix representations. This is a desirable situation because, in most instances, it is easier to work with matrices than to work with abstract groups.[68]

6. If there is an irreducible representation, then there is a matrix algebra.

7. If there is a matrix algebra, then there is a manageable calculation.

   Mathematicians are masters at matrix algebra.

8. If there is a manageable calculation, then there is a manageable mathematical relationship.

---

[67]This is basically what slide rules do for those of us who persist in using slide rules.

[68]Irreducible representations, also known as irreps, are isomorphisms to symmetry groups wherein the group elements are matrices and the group operations (i.e., the symmetry transformations) are matrix operators. Irreducible representations are representations that cannot be decomposed into nontrivial sub-representations.

Empirically, most physical relationships (in either the macroscopic world or the quantum world) can be expressed in short equations.

9. If there is a mathematical relationship, then there is a general physical reality expressed by the relationship.

4.23 Proving the physical universe is made of mathematics

"Mathematics is the part of science you could continue to do if you woke up tomorrow and discovered the universe was gone".
—Dave Rusin

Each of us creates for ourselves a subjective understanding of reality. For most of us, this sense of reality has nothing to do with mathematics. Mathematics, we are taught, is little more than a tool for solving a restricted set of technical problems. Personal proficiency in mathematics is an option, not a requirement. This being the case, it is difficult to accept that **mathematics is equivalent to reality**. Let's not be dismissive. When evaluated objectively, there does not seem to be much difference between these two sublime abstractions.

- Mathematics is timeless.

  Mathematics has no beginning and no ending, and this fits with our intuitive sense that existence always exists.

- Mathematics is fully descriptive of reality.

  We think of reality as spacetime and the forces and particles operating in spacetime. All of this is described mathematically, and there does not seem to be anything we can observe that cannot be described mathematically.

- Mathematics, like reality, is inviolate.

  Gödel's incompleteness theorem tells us that there will always be true statements that cannot be proven. Does Gödel's incompleteness theorem apply to all of reality? Yes. Suppose there existed an all-powerful being who had created the universe and all of the laws that govern reality. We might believe that it would be within the power of an all-powerful being to prove any true statement that happens to be true. Not so. No power can violate Gödel's incompleteness theorem. The math won't budge.

- Mathematics provides an isomorphism of reality

  Max Tegmark has made the argument that mathematics is isomorphic to reality, meaning that there is a 1 to 1 mapping between reality and its mathematical description, that preserves relationships (i.e., the mathematical construct is equivalent to the mathematical laws of physical reality). If two structures are isomorphic, he reasons, then there is no distinction between them. Therefore the isomorphism between the mathematical structure and reality establishes that they are both one and the same. Hence, reality is a mathematical structure.[69]

- Other than mathematics, there seems to be no alternative way to explain what we think we know about reality.

  Let's review a few assertions about the nature of reality that require mathematical explanation.

  - Matter is composed of point particles (better known as fundamental particles) that have no size and no precisely discernible location. The point particles are not composed of "stuff". They are made of nothing but their properties. When we encounter large, non-fundamental particles, such as protons and neutrons, we find that they are composed of point particles and the forces that bind them. That is to say that protons and neutrons have the same size as electrons: zero. What we imagine to be the size of a proton or a neutron corresponds to the extent of the containment field (that holds their constituent quarks and gluons), and the mass of the proton or neutron is basically the mass-equivalent of the containment field's energy. That is to say that the mass of the proton, which is nearly two thousand times the mass of the electron, is produced by the confinement energy (of the strong force) generated by gluon-quark interactions.
  - We can only understand matter, force, and energy in terms of their properties. That is to say, everything that "happens" is just an interplay of movement and forces, in turn consisting of their mathematical descriptions.[70]
  - The universal determinates of the relations that govern the universe are dimensionless constants. It is difficult to imagine how a small coterie of numbers could have such sway in anything other than in a purely mathematical reality.
  - Symmetries account for the laws of the universe, but the symmetries are intangibles having only mathematical meaning. The gauge symmetries, in particular, seem to be creatures composed entirely of mathematics. It is difficult to perceive of the symmetries of nature as anything other than mathematics reified (i.e., made real through application). [71]

---

[69]Max Tegmark is a proponent of the mathematical theory of everything. For an explanation of this theory, read his article, titled "The Mathematical universe," in: Foundations of Physics 38:101-150, 2008.

[70]Paraphrasing the mathematician and philosopher Bertrand Russell, we do not really know what matter is. We only know the mathematical description of the properties of matter (see Russell B. *The Analysis of Matter*, Kegan Paul publishers, 1927).

[71]Where there is a symmetry there is a field; where there is a field, there are conserved quantities; where there are conserved quantities there are quanta; where there are quanta, there are

Our deepest understanding of physical existence can be expressed as mathematical relationships. Nonetheless, we are reluctant to equate reality with mathematics. Perhaps our hesitation stems from the common belief that our existence and our destiny are determined by powers that lie beyond human comprehension. It's fun to prove things, but for the moment, **our best clues to an understanding of reality lies within our puzzlements, not our convictions.**

particles. Reality emerges from the mix. If we were to remove symmetries, then our universe would vanish, and the sales of this book would plummet.

5

Handy Math

"If you want to learn about Nature, to appreciate Nature, it is necessary to understand the language that she speaks in".
—Richard Feynman, referring to the study of mathematics

It is impossible to fully appreciate the laws of nature without knowing a bit of mathematics. In this chapter, some of the most useful formulas of physics, many of which have already appeared in prior chapters, are proven. In every case, the proofs are designed to be short and simple, occasionally sacrificing some degree of rigor for the sake of accessibility. Some of the topics for proof include integration by parts, rules for the natural logarithms, trigonometric laws, the principle of mathematical induction, Green's theorem, Bayes' theorem, the Fourier transformation, convolution, the Dirac impulse function, theorems for groups, properties of even and odd functions, the Taylor expansion, Stirling's approximation, the value of the Gaussian integral, and Laplace's method.

DOI: 10.1201/9781003516378-5

5.1 Proving the Pythagorean theorem

The Pythagorean theorem states that for any right triangle, the square of the hypotenuse (long side, or side opposite to the right angle), is equal to the sum of the squares of the other two sides. There are literally hundreds of clever proofs for the Pythagorean theorem, and here is Albert Einstein's offering.[1]

Consider the following construction, wherein a right angle project extends from corner C to side c.

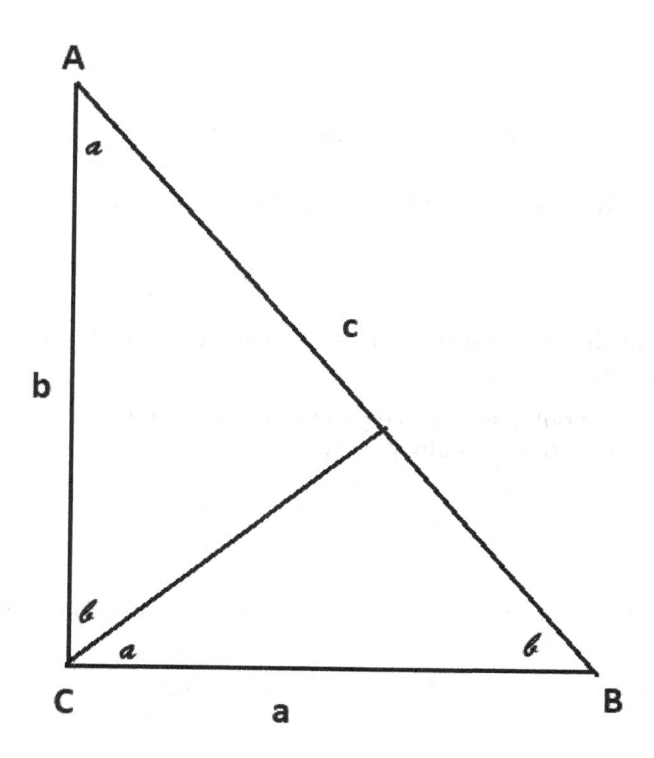

FIGURE 5.1: Graphic for Einstein's proof of the Pythagorean theorem

Using simple trigonometric relationships, we can evaluate the hypotenuse, c.

[1]Pythagoras lived about 2,500 years ago, but the same theorem was probably known to the Egyptians about 3,500 years ago.

$$c = b\sin\beta + a\sin\alpha$$

But

$$sin\alpha = \frac{a}{c}$$

and

$$sin\beta = \frac{b}{c}$$

So,

$$c = b\sin\beta + a\sin\alpha = b(\frac{b}{c}) + a(\frac{a}{c})$$

multiplying both sides by c yields the Pythagorean theorem.

$$c^2 = b^2 + a^2$$

For a proof that you are likely to find in standard textbooks, consider the following construction:

The area of the outer square is equal to the area of the inner square plus the areas of the four triangles along its edges.

$$(a+b)^2 = c^2 + 4\frac{ab}{2}$$

Expanding $(a+b)^2$ and evaluating $4\frac{ab}{2}$, and rearranging, gives us:

$$c^2 = a^2 + b^2 + 2ab - 2ab$$

or

$$c^2 = a^2 + b^2$$

Thus proving that c^2, the hypotenuse of any of the 4 right triangles, is equal to $a^2 + b^2$, the sum of the squares of the other two sides.

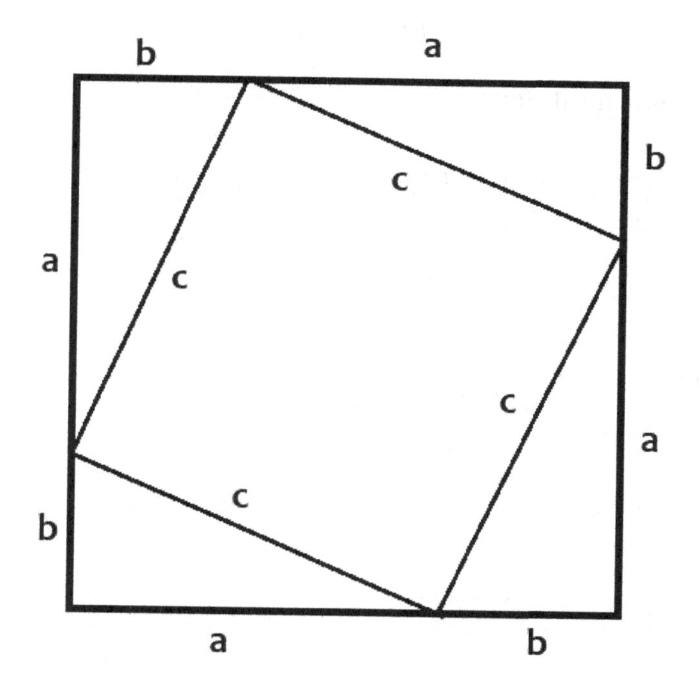

FIGURE 5.2: Graphic for a second proof of the Pythagorean theorem

5.2 Proving $e^{(ln(y))} = y$

Let

$$e^{(ln(y))} = x$$

Take the natural logarithm of both sides:

$$ln(e^{(ln(y))}) = ln(x)$$

The left side is equivalent to:

$$ln(y) \times ln(e^1) = ln(y) \times ln(e) = ln(y) \times 1 = ln(y)$$

so,

$$ln(y) = ln(x)$$

and therefore,

$$y = x = e^{(ln(y))}$$

5.3 Proving that $x^n = a^{n(\log_a x)}$

By definition

$$y = \log_a x \text{ iff } x = a^y$$

Let

$$x = a^y$$

then

$$x = a^{\log_a x}$$

and

$$x^n = a^{n(\log_a x)}$$

5.4 Proving $log_a x = log_c x / log_c a$

By definition $y = log_a x$ iff $x = a^y$

Let's take the log_c of both sides of the exponential expression

$$log_c x = log_c a^y$$

Substituting $y = log_a x$ for y,

$$log_c x = log_c a^{log_a x}$$

so,

$$log_c x = log_a x \times log_c a$$

or

$$log_a x = \frac{log_c x}{log_c a}$$

5.5 Proving $log_a c = 1/log_c a$

We have already proven that $log_a x = log_c x/log_c a$.[2] Let's substitute c for x. Then,

$$log_a c = \frac{log_c c}{log_c a}$$

But, we know that $log_c c = 1$. So,

$$log_a c = \frac{1}{log_c a}$$

[2]See section titled "Proving $log_a x = log_c x/log_c a$"

5.6 Proving the chain rule

The chain rule tells us that if $y = f(x)$ and $x = g(t)$, then the derivative of y with respect to t is:

$$\frac{dy}{dt} = \frac{dy}{dx} \times \frac{dx}{dt}$$

There are several available proofs of the chain rule, but we'll discuss the shortest.

1. If $x = g(t)$, then there is infinitesimal Δt such that

$$\Delta x = g(t + \Delta t) - g(t)$$

2. In the same way, for $y = f(x)$, there is infinitesimal Δx such that

$$\Delta y = f(x + \Delta x) - f(x)$$

3. But $\frac{\Delta y}{\Delta t} = \frac{\Delta y}{\Delta t} \cdot \frac{\Delta x}{\Delta x}$
4. Rearranging, $\frac{\Delta y}{\Delta t} = \frac{\Delta y}{\Delta x} \cdot \frac{\Delta x}{\Delta t}$
5. Applying the infinitesimal for each Δ value

$$\frac{dy}{dt} = \frac{dy}{dx} \cdot \frac{dx}{dt} \qquad \text{The chain rule}$$

5.7 Proving the derivative of the natural log of y is $\frac{1}{y}$

First, we can always say that

$$\frac{dy}{dy} = 1$$

In an earlier proof, we established that

$$y = e^{(ln(y))}$$

Substituting back, we have:

$$\frac{dy}{dy} = \frac{d}{dy}e^{(ln(y))} = 1$$

Now, let's use the chain rule:

$$\frac{d}{dx}g(h(x)) = \frac{dg}{dh} \times \frac{dh}{dx}$$

In this case, h(x) is ln(y) and g(x) is $e^{(ln(y))}$

So, because the derivative of a natural log is the natural log,

$$1 = \frac{d}{dy}e^{ln(y)} \times \frac{d}{dy}ln(y) = e^{ln(y)} \times \frac{d}{dy}ln(y) = y \times \frac{d}{dy}ln(y)$$

But $e^{ln(y)} = y$, so by dividing both sides of the equation by y, we reach our proof:

$$\frac{1}{y} = \frac{d(ln(y))}{dy}$$

5.8 Proving $\int \frac{1}{y}\,dy = ln(y)$

In section titled "Proving the derivative of the natural log of y is $\frac{1}{y}$", we proved that

$$\frac{1}{y} = \frac{d(ln(y))}{dy}$$

Let's take the integral of both sides of the equation.

$$\int \frac{1}{y} dy = \int d(ln(y))$$

The fundamental theorem of calculus tells us that the integral of a derivative of a function is the function itself, so the right side of the equation is just the function (ln(y), and,

$$\int \frac{1}{y}\,dy = ln(y)$$

5.9 Proving the area under a function is $\int_a^b f(x)dx$

We need to show that for a continuous function, the area under the curve (representing the function) over an interval, is the integral of the difference between the integral evaluated at the beginning of the interval and the and the integral evaluated at the end of the interval.

The proof consists of nothing more than an examination of the meaning of an integral of a function over an interval. The integral is simply the summation of the value of the function at a small interval times the width of the small interval, yielding the area of the small rectangle rectangle under the function. Adding up all of the small areas yields the total area under the function.[3]

[3]Students should use caution when applying this rule to functions whose coordinates cross into negative ranges. Computed areas may have negative values, but physical areas are always positive, and results must be converted to absolute values.

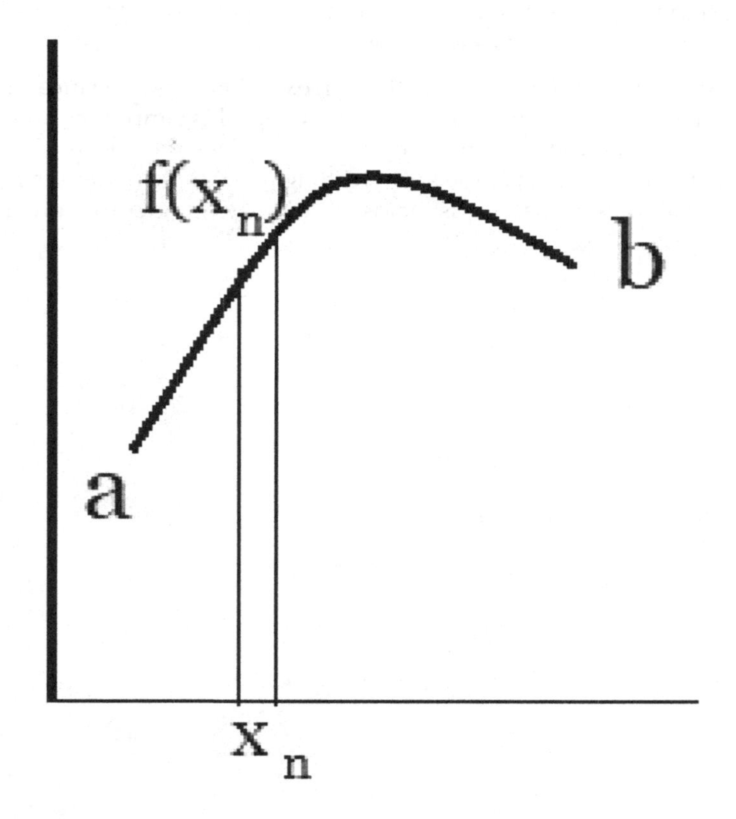

FIGURE 5.3: Every small interval in the curve encompasses a rectangular area equal to the value of the function at the interval times the width of the interval. Adding the areas of all of the small rectangles, we get the total area under the function in the full interval, a to b.

5.10 Proving a function's average value is $(1/b-a)\int_a^b f(x)dx$

We need to prove that the average value of a continuous function equals the integral of the function divided by its interval of integration. Stated as a mathematical equation:

$$f_{average} = \frac{1}{b-a}\int_a^b f(x)dx$$

We'll construct a proof based on probabilistic sampling of the continuous function, f(x) over the integral $a \to b$.

1. Let's randomly determine the falue of f(x) at n points in the interval $a \to b$
2. The average value of the function is the sum of all the values divided by n, the number of values sampled. This average becomes more accurate as n grows larger (i.e., as n approaches infinity).
3. Regardless of how large n becomes, we have $n = \frac{b-a}{\Delta x}$, where Δx is the average interval between samples.
4. So, the average value of f(x) is the sum of all the values divided by $b - a/\Delta x$. This is

$$f_{average} = \sum_{i=1}^{n} \frac{f(x_i)}{\frac{b-a}{\Delta x}}$$

   Which can be rewritten as,

5.

$$f_{average} = \frac{1}{b-a}\sum_{i=1}^{n} f(x_i)\Delta x$$

6. But $\sum_{i=1}^{n} f(x_i)\Delta x$ is the integral of f(x), as $n \to \infty$ in the integral $a \to b$
7. So we have shown that

$$f_{average} = \frac{1}{b-a}\int_a^b f(x)dx$$

This relationship is particularly useful for the following two reasons:

- Whenever we know the average value of a continuous function within an interval, we can easily compute its integral directly, without solving the integral formula.
- We can compute the integral by imitating the integration process, using the random number generator available on all computers.

The area under the curve of a function is the integral of the function over the range of x. Depending on the function, the integral may be very hard to compute. In these cases, calculation using a random number generator can be quick and easy. All we need to remember is that the area under the curve of a function is also equal to the x-interval over which the function is evaluated multiplied by the average value of the function within the interval. Let's see how we can use this information to compute the area under a curve, using a simple random number generator. Let's calculate the area under the curve for the function $f(x) = x^3 - 1$. We can use a very simple Python script, (vida infra), that can be modified to calculate almost any integral (of a continuous function), over any interval.

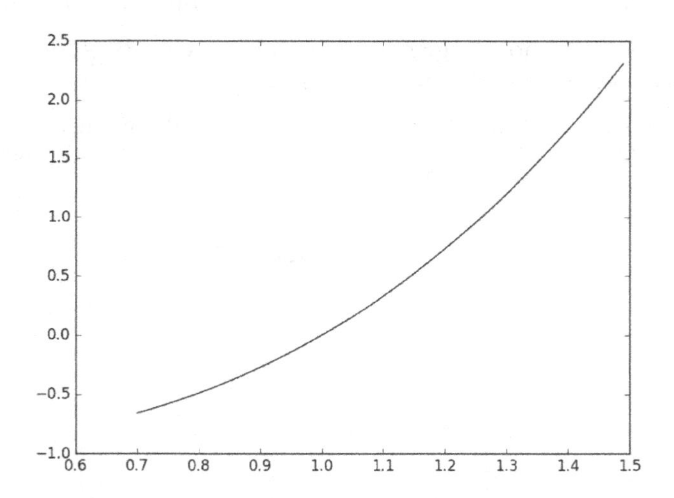

FIGURE 5.4: The plot of the function $x^3 - 1$, in the interval 0.7 to 1.5.

Let's use a random number generator, in a Python script to evaluate the integral.

```
import math, random, itertools
def f(x):
  return (x*x*x -1)
range = 2          #evaluating the integral from x=1 to x=3
```

```
running_total = 0
for i in itertools.repeat(None, 1000000):
  x = float(random.uniform(1,3))
  running_total = running_total + f(x)
integral = (running_total / 1000000) * range
print(str(integral))
```

Here is the output:

```
18.001426696174935
```

The output comes very close to the exact integral, 18, for the interval between 1 and 3. How did our short Python script calculate the integral? The integral of a function is equivalent to the area under the curve of the function, in the chosen interval. The area under the cover is equal to the size of the interval (i.e., the x coordinate range) multiplied by the average value of the function in the interval. By calculating a million values of the function, from randomly chosen values of x in the selected interval, and taking the average of all of those values, we get the average value of the function. Just as long as we can successfully calculate values of the function, from randomly chosen values of x, within a selected interval, we can calculate the area under the curve When we multiply this by the interval, we get the area, which is also equal to the value of the integral.

5.11 Proving the fundamental theorem of calculus

The fundamental theorem of calculus tells us that the integral and the derivatives of a continuous function are inverses of one another. Another way of stating this is that the integral of the derivative of a function yields the original function.

Put into the form of an equation,

$$f(x) = \int_0^x f'(t)dt$$

This is equivalent to saying that for some function F(x) that is equal to the integral of some other function, f(t), we can infer that the differential of the function, F′(x) is equal to the other function (i.e., $F(x) = \int_a^x f(t)\,dt$ tells us that $F'(x) = f(x)$).

We'll settle for a non-rigorous proof that extrapolates from a single interval in a continuous function.

From the definition of an integral,

$$F(x + \Delta x) - F(x) = \int_x^{x+\Delta x} f(t)\,dt,$$

In the small interval, Δx, the integral of $f(t)$ is simply a value of f(t) in the interval (which we will call f(c) times the width of the interval.

$$F(x + \Delta x) - F(x) = f(c)\Delta x,$$

Taking the limit as $\Delta x \to 0$,

$$\lim_{\Delta x \to 0} \frac{F(x + \Delta x) - F(x)}{\Delta x} = \lim_{\Delta x \to 0} f(c)$$

As $\Delta x \to 0$, $f(c) \to f(x)$, which is just another way of saying,

$$F'(x) = f(x)$$

Which is what we needed to prove.

5.12 Proving the mean value theorem for integrals

The mean value theorem for integrals is closely related to our finding in the prior section, that the average value of a function is equal to $= \frac{1}{b-a} \int_a^b f(x)dx$. The mean value theorem tells us that for a continuous function over an interval, the the average of the function lies somewhere on the curve (i.e., the average must be expressible by the function).

Stated formally, If f(x) is continuous in the interval [a,b], then there is a number c in the interval such that:

$$\int_a^b f(x)dx = f(c)(b-a)$$

This theorem is proven directly from the two preceding theorems.

We have previously proven that for a continuous function over an interval, the average value of the function is $= \frac{1}{b-a} \int_a^b f(x)dx$. Rewritten:

$$f_{avg}(b-a) = \int_a^b f(x)dx$$

We know that the integral of the function can be expressed as the difference between F(a) and F(b), where $F(x) = \int_a^x f(t)dt$. Let's integrate AND differentiate the left side of the equation:

$$f_{avg}(b-a) = (F(b) - F(a))'$$

The fundamental theorem of calculus tells us that for continuous functions, the differential of the integral of a function at a point is the same as the function at that point (see prior section). Therefore, (F(b) - F(a))′ must have a value that can be expressed as the value of the function between a and b. Therefore, the average value of the function between a and b (f_{avg}) must be expressible as a value of the function in the interval.

5.13 Proving the product rule

Here is the product rule:

$$\frac{d(f(x)g(x))}{dx} = \left(\frac{df(x)}{dx}\right)g(x) + f(x)\left(\frac{dg(x)}{dx}\right)$$

Expressed alternately,

$$(fg)' = f'g + g'f$$

There are several simple proofs for the product rule, but one of the simplest involves using the previously proven equation for the derivative of the natural logarithm, namely:

$$\frac{1}{y} = \frac{d(ln(y))}{dy}$$

Or,

$$\left(\frac{1}{y}\right)dy = d(ln(y))$$

Or, in an even simpler form,

$$\frac{y'}{y} = (ln(y))'$$

Let's let y equal the product of f(x) and g(x)

$$y = f(x)g(x)$$

Or, in simple form,

$$y = fg$$

Substituting for y,

$$\frac{(fg)'}{fg} = (ln(fg))' = ln(f)' + ln(g)'$$

Multiplying both sides by fg, and remembering that $\frac{y'}{y} = (ln(y))'$, we substitute again,

$$(fg)' = (fg)(\frac{f'}{f} + \frac{g'}{g})$$

And,

$$(fg)' = f'g + g'f \qquad \text{The product rule}$$

Directly from the product rule, we can apply implicit differentiation (i.e. differentiating on functions that operate on both x and y), to derive many useful relationships, such as:

$$\begin{aligned}
\frac{d}{dx}(y^2) &= 2y\frac{dy}{dx} \\
\frac{d}{dx}(siny) &= cosy\frac{dy}{dx} \\
\frac{d}{dx}(tan^{-1}y) &= \frac{1}{1+y^2} \times \frac{dy}{dx}
\end{aligned}$$

5.14 Proving the quotient rule: $(f/g)' = (f'g - fg')/g^2)$

We shall prove that for two functions, f(x) and g(x): $(f(x)/g(x))' = (f(x)'g(x) - f(x)g(x)')/g(x)^2)$

Let's substitute y for $f(x)/g(x)$. Then

$$lny = ln(f(x)/g(x)) = lnf(x) - lng(x)$$

And now let's take the derivative of both sides of the equation

$$(lny)' = (lnf(x))' - (lng(x))'$$

In section titled "Proving the derivative of the natural log of y is $\frac{1}{y}$", we proved,

$$\frac{1}{y} = \frac{d(ln(y))}{dy}$$

Or, multiplying both sides by dy (i.e., y'):

$$y'/y = (ln(y))'$$

Applying, the derivatives of the logarithmic terms on both sides of the quotient equation:

$$y'/y = f'(x)/f(x) - g'(x)/g(x)$$

Or,

$$y' = y(f'(x)/f(x) - g'(x)/g(x))$$

Or,

$$y' = (f(x)/g(x))' = \frac{f(x)}{g(x)}\left(\frac{f'(x)}{f(x)} - \frac{g'(x)}{g(x)}\right)$$

Which, in abbreviated notation, reduces to,

$$(f(x)/g(x))' = (f(x)'g(x) - f(x)g(x)')/g(x)^2) \qquad \text{The quotient rule}$$

5.15 Proving integration by parts

Integration by parts is the one-dimensional equivalent of the divergence theorem, which is the mathematical equivalent of the continuity equation of physics, which is the general expression that accounts for the bulk of Maxwell's laws, which are expressions of the conservation laws of quantum physics, which are the result of universal symmetries, which embody all of the fundamental properties of reality. So, it is probably worth our while to take a moment or two to prove integration by parts.[4]

Integration by parts states that,

$$\int u dv = uv - \int v du$$

The proof of integration by parts comes directly from the product rule (see section titled "Proving the product rule"). In essence, the product rule does for differentiation what integration by parts does for integration.

Applying the product rule to two functions of x,

$$\frac{d(uv)}{dx} = v\frac{du}{dx} + u\frac{dv}{dx}$$

Integrating both sides with respect to x we have:

$$\int d(uv)dx = \int v\frac{du}{dx}dx + \int u\frac{dv}{dx}dx$$

But the integral on the left side is just the anti-derivative of the derivative of uv, so:

$$uv = \int v du + \int u dv$$

Or

$$\int u dv = uv - \int v du \qquad \text{Integration by parts}$$

[4]Here is an amusing story (possibly apocryphal), recounted by Willie Wong. When Peter Lax was awarded the National Medal of Science, the other recipients (presumably non-mathematicians) asked him what he did to deserve the Medal, Lax responded, "I integrated by parts". From a note written by Willie Wong, Apr 24, 2011, on StackExchange.

In the past several sections, we have proven the rules of differentiation that we use for the types of equations that appear throughout the physical sciences. Scientists need to know how one variable changes in relation to some other changing variable. Many such problems involve the relationship between two variables, written as x and y. The general technique for solving these equations is known as **"separation of variables"**. It involves the following simple steps:

1. Use our proven rules to rewrite the differential equation with the x and dx terms on one side of the equation and the y and dy terms on the other side.
2. Integrate both sides of the equation to get rid of the differentials (i.e., dx and dy)
3. The resulting equation will be an expression of x and y and some constant.
4. Evaluate the initial conditions of the physical model to determine the value of the constant.
5. Submit the final equation to the Nobel prize committee.

5.16 Proving a function inverse yields the function's integral

Let's start integrating the continuously differentiable function, f(x), using integration by parts.

$$\int f(x)dx = xf(x) - \int xf'(x)dx$$

Using the definition of an inverse function, we know,

$$f^{-1}(f(x)) = x$$

Let's substitute $f^{-1}(f(x))$ for x back into the right-side integral from the first equation.

$$\int f(x)dx = x(f(x)) - \int f^{-1}(f(x))f'(x)dx$$

Now, let's substitute u for $f(x)$ in the right-side integral.

$$\int f(x)dx = xf(x) - \int f^{-1}(u)du$$

This tells us that to find the integral of a function, all we need to know is the value of the function and the integral of its inverse. As it happens, there are many instances in quantum physics when the inverse of a function and the integral of the inverse are known.

5.17 Proving area of a circle $= \pi r^2$

The simplest proof is by integration. From the center of a circle, we move a short distance, *dr*, from and determine the area of a small strip of the circle that can be straighted into a rectangle of width *dr* and of width $2\pi r$, and we repeat in increments of *dr* up to the full radius of the circle, *R*.

$$A = \int_0^R 2\pi r dr = \left| 2\pi \frac{1}{2} r^2 \right|_0^R = \pi R^2$$

We could achieve the same result, without integral calculus, by building a triangle from unwound strips of concentric circles.

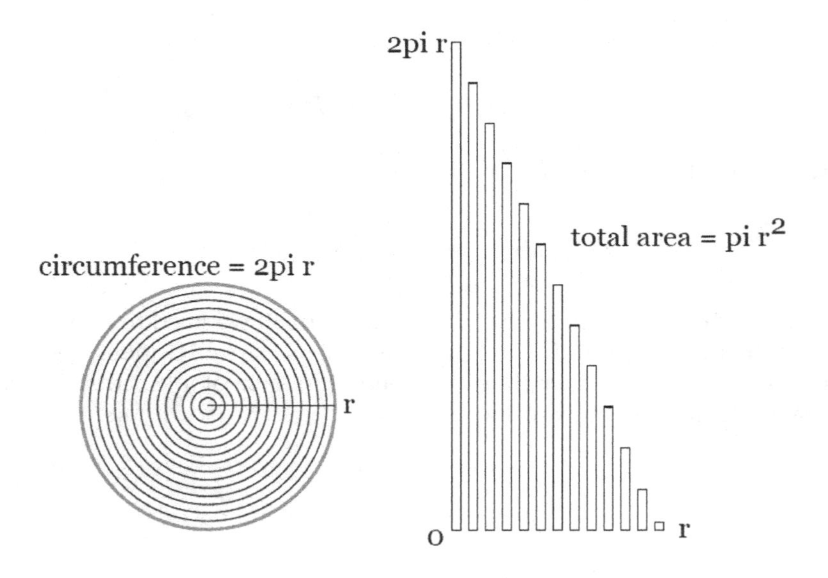

FIGURE 5.5: Concentric rings of a circle are unwrapped and laid out as straight rectangles.

The area of the triangle is equal to the area of the circle (because the triangle is composed of the concentric rings of the circle). The area of the triangle is equal to 1/2 the base (r) times the height ($2\pi r$), or πr^2.

5.18 Proving the value of π using a random number generator

There are hundreds of ways in which a reasonable value for π can be determined. To demonstrate how random number generators can simulate physical operations, we will calculate the value of pi, without measuring anything, and without resorting to summing an infinite series of numbers. Here is a simple python script that does the job.

```
import random, itertools
from math import sqrt
totr = 0; totsq = 0
for i in range(10000000):
  x= random.uniform(0,1)
  y= random.uniform(0,1)
  r= sqrt((x*x) + (y*y))
  if r < 1:
    totr = totr + 1
  totsq = totsq + 1
print(float(totr)*4.0/float(totsq))
```

We can easily plot the results. The number of randomly generated coordinates that fall within the circle divided by the total number of randomly generated coordinates must equal the area of a quarter circle divided by the area of the square. We multiply by four to find the area of the full circle. Or,

$$\pi r^2 = 4 \times \text{number of points in circle/total number of points}$$

In this case, $r = 1$, and the equation evaluates to π. When we generate 10,000,000 coordinates, we find that the value of *pi* is approximated as 3.1416816.

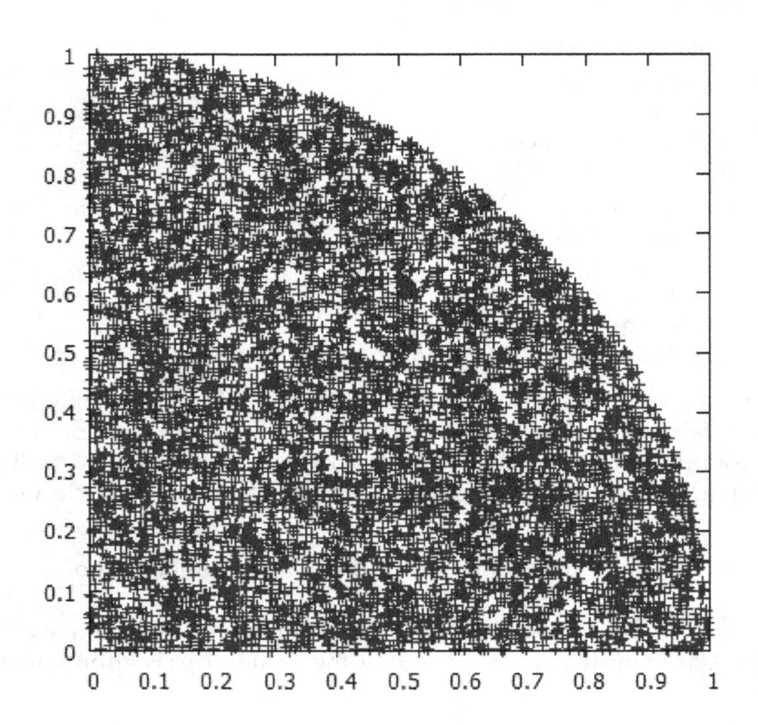

FIGURE 5.6: Random points filling the area of a quarter circle

5.19 Proving Euler's formula and Euler's identity

Imagine a function, $f(\theta)$, with the following property:

$$f(\theta) = e^{-i\theta}(\cos\theta + i\sin\theta)$$

We'll use the product rule (see section titled "Proving the product rule") to differentiate $f(\theta)$.

$$f'(\theta) = e^{-i\theta}(i\cos\theta - \sin\theta) - ie^{-i\theta}(\cos\theta + i\sin\theta)$$

Evaluating and combining terms,

$$f'(\theta) = e^{-i\theta}(i\cos\theta - \sin\theta - i\cos\theta + \sin\theta)$$

In the equation above, we see that everything on the left cancels out, indicating that $f'(\theta) = 0$, which tells us that $f(\theta)$ must be a constant.

We can determine the constant value of $f(\theta)$ by substituting any value of θ into the original equation (that's one of the perks of being a constant). Let's substitute zero as our value of θ.

$$f(0) = e^{-i0}(\cos 0 + i\sin 0) = 1$$

Evaluating the left-hand side of the equation, we find that $f(\theta)$ always evaluates to 1. Thus,

$$1 = e^{-i\theta}(\cos\theta + i\sin\theta)$$

Dividing both sides by $e^{-i\theta}$,

$$e^{i\theta} = \cos\theta + i\sin\theta \qquad \text{Euler formula}$$

When we substitute π as θ in Euler's formula, we have,

$$e^{i\pi} = \cos\pi + i\sin\pi$$

Evaluating, we get,

$$e^{i\pi} = -1 + i0$$

Rearranging,

$$e^{i\pi} + 1 = 0 \qquad \text{Euler's identity}$$

Euler's identity, widely credited as the most beautiful equation in mathematics, manages to include virtually every important mathematical symbol in one compact relationship $(e, i, \pi, 1, 0, and =)$.

Now that we have Euler's formula, we can derive a few useful variations. Substituting kx for x,

$$e^{ikx} = cos(kx) + isin(kx)$$

The substitution of kx for x is equivalent to raising e^{ix} to the kth power.

$$e^{ikx} = (e^{ix})^k = (cos(x) + isin(x))^k = cos(kx) + isin(kx)$$

This latter equivalence is known as the DeMoivre identity, discussed more fully in the section titled "Proving the DeMoivre identity".

5.20 Proving any complex number can be expressed in terms of Euler's formula

Complex numbers have the form $z = x + iy$, where z,x, and y are real numbers. We can visualize this relationship on a complex coordinate plane, as shown:

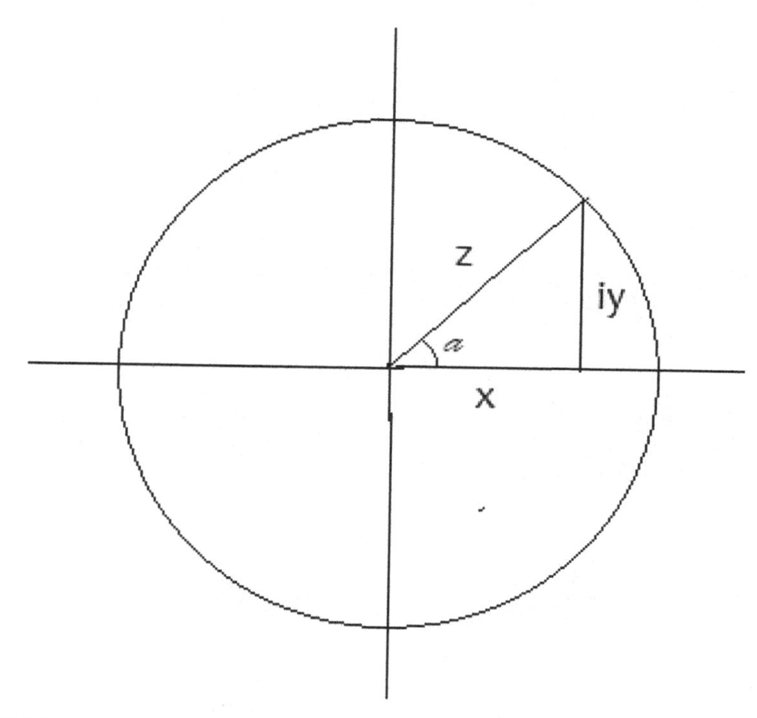

FIGURE 5.7: A complex number, $x + iy$, represented on the complex plane, with z being the hypotenuse and alpha being the angle at the origin.

We see that,

$$z^2 = x^2 + |(iy)^2|$$

We also see that,

$$x = zcos\alpha \qquad iy = zi\,sin\alpha$$

Euler's formula tells us,

$$e^{i\alpha} = cos\alpha + i\,sin\alpha$$

Or, in this case,

$$ze^{i\alpha} = zcos\alpha + zi\,sin\alpha = x + iy = z$$

Hence, any complex number can be represented by Euler's formula.

We can also show that complex numbers, in polar form, can easily be expressed in terms of Euler's formula. Any complex number $(a + bi)$, can be expressed in polar coordinates, with $a = r(\cos\theta)$ and $ib = r(i\sin\theta)$.

Together, $a + bi$, in polar coordinates is $r(\cos\theta + i\sin\theta)$, or, as per Euler's formula, $re^{i\theta}$

5.21 Proving a complex number multiplication rule

Suppose we want to multiply two complex numbers, a and w where u=a+bi and w=c+di

We'll rewrite u and w, using the hypotenuse in the complex plane of u, which we'll call y and the hypotenuse in the complex plane of w that we'll call z. The angles of u and w are alpha and beta, respectively

$$u = a + bi = y(cos\alpha) + iy(sin\alpha)$$

$$w = c + di = zcos\beta + iz\,sin\beta$$

When we multiply the two complex numbers, we get:

$$u \cdot w = (y(cos\alpha) + iy(sin\alpha)) \cdot (z(cos\beta) + iz(sin\beta))$$

Or,

$$uw = (ycos\alpha)(zcos\beta) + (ycos\alpha)(zi\,sin\beta) + (yi\,sin\alpha)(zcos\beta) + (yi\,sin\alpha)(zi\,sin\beta)$$

Or,

$$uw = yz(cos\alpha)(cos\beta) + yzi\,((cos\alpha)(sin\beta) + (sin\alpha)(cos\beta)) - yz(sin\alpha)(sin\beta)$$

But from the addition formulas for sines and cosines, we know that:

$$(cos\alpha)(cos\beta) - (sin\alpha)(sin\beta) = cos(\alpha + \beta)$$

And,

$$(cos\alpha)(sin\beta) + (sin\alpha)(cos\beta) = sin(\alpha + \beta)$$

Therefore,

$$uw = yz(cos(\alpha + \beta) + i\,sin(\alpha + \beta)) = yz \cdot e^{i(\alpha+\beta)}$$

Therefore, when we multiply two complex numbers, their lengths (y and z, in this case) are multiplied, and their angles (α and β) are added to the exponential term.

5.22 Proving the DeMoivre identity

Here's the DeMoivre identity.

$$(\cos\theta + i\sin\theta)^n = \cos(n\theta) + i\sin(n\theta)$$

The proof follows easily from Euler's formula, which states,

$$e^{i\theta} = \cos\theta + i\sin\theta$$

Let's raise both sides of Euler's formula to the nth power.

$$(e^{i\theta})^n = (\cos\theta + i\sin\theta)^n$$

The left-hand side can be represented as,

$$(e^{i\theta})^n = e^{i(n\theta)}$$

So, we can just substitute $n\theta$ as our angle in Euler's formula, to yield the DeMoivre identity.[5]

$$e^{i(n\theta)} = \cos(n\theta) + i\sin(n\theta) = (\cos\theta + i\sin\theta)^n$$

[5] The DeMoivre identity can give us the nth roots of unity. Let's set $x = \frac{2\pi}{n}$. Then, $(\cos\frac{2\pi}{n} + i\sin\frac{2\pi}{n})^n = \cos 2\pi + i\sin 2\pi = 1$. The nth roots of unity form an irreducible representation of any cyclic group of order n. DeMoivre's simple and unassuming identity has significance in quantum physics, where spatial symmetries of particles have irreducible representations.

5.23 Proving the uniqueness of inverse elements of groups

"Numbers measure size, groups measure symmetry".
—Mark A. Armstrong

Let's quickly review the definition of a group:

- A group is a set having the properties of closure and associativity under an operation; having an identity element; and having an inverse element for every member of the set.
- Specifically, for some operation (which we'll call "x" in this example, "closure" tells us that for all a, b in G, the result of the operation, a x b, is also in G.
- Associativity tells us that for all a, b and c in G, (a x b) x c = a x (b x c).
- To have an identity element, there must exist an element e in G such that, for every element a in G, the equation e x a = a x e = a holds. The identity element is unique within a group, and we speak of "the" identity element.
- To have an inverse element, there must exist for each a in G, an element b in G, such that a x b = b x a = e, where e is the identity element.
- A simple group is a group G that does not have any normal subgroups except for the trivial group and G itself.

For emphasis, here are the associative and identity properties of any group without specifying the operation (other than referring to the operation with a dot)

$$a \cdot (b \cdot c) = (a \cdot b) \cdot c$$

And,

$$a \cdot a_{inverse} = I = a_{inverse} \cdot a$$

We'll prove that the inverse of every group element is unique

Suppose that a group element p has two different inverses, m and n. We'll call the identity element of the group q.

By the definition of the group identity (q), we know that,

$$m = m \cdot q$$

By the definition of an inverse, we know that the identity element is equal to p times either of its inverse elements (m and n). So,

$$m = m \cdot (p \cdot n)$$

Every group has the association property. So:

$$m = (m \cdot p) \cdot n$$

But we know that p times its inverse must equal the identity element, q

$$m = q \cdot n$$

And any element multiplied by the identity element is, by definition, itself.

$$m = n$$

Therefore the two inverse elements of the group element p (i.e., m and n), must be identical to one another.[6]

[6]For physicists, the most interesting type of group is the symmetry group, constituting the set of all symmetry transformations (i.e., transformations that retain the properties of the object being transformed). The elements of the group are its transformation operations. When the group operations are symmetries, then we say that the transformations are symmetry transformations. All of the groups that represent the physical operations of the universe are symmetry groups.

5.24 Proving the identity element of a group is unique

The prior proof provides everything we need to prove that the identity element of any group is unique.

1. Every element of a group has an inverse.
2. Therefore, the identity element, being an element of a group, must have an inverse.
3. The inverse of the identity element is the identity element
4. We have already proven that the inverse of an element is unique

   Therefore,

5. The identity element (which is an inverse) is unique

5.25 Proving irrational numbers exist

To prove that irrational numbers exist, we only need to find one number that we can prove to be irrational.

As it happens, it is rather easy to prove that $\sqrt{2}$ is irrational.

1. Let's start by assuming the contrary (i.e., that $\sqrt{2}$ is rational).
2. If $\sqrt{2}$ is rational, then it can be expressed as the ratio of two integers, a/b, where $b \neq 0$.
3. Let's imagine that whatever a/b is, it has been expressed in its lowest terms (any integer fraction has the lowest term).
4. Then, by squaring both sides of the equation, $2 = a^2/b^2$.
5. Therefore, $a^2 = b^2 \times 2$, which means that a^2 is an even number.
6. Therefore, by definition of an even number, $a = 2c$, where c is an integer.
7. Returning to step 5, $(2c)^2 = 2b^2$ or $4c^2 = 2b^2$ and $b^2 = 2c^2$.
8. Therefore, b^2 is an even number.
9. Because only an even number squared yields another even number, we know that b must be an even number.
10. Because both a and b are even numbers, then a/b is not the lowest term fraction.
11. Therefore, our initial assumption must be wrong, and $\sqrt{2}$ must be irrational.

This is an example of a proof by contradiction. It is one of many techniques that mathematicians use to prove existence.[7]

[7]This example was adapted from a proof in *Mathematical Proofs: A Transition to Advanced Mathematics* by Gary Chartrand, Albert D. Polimeni, and Ping Zhang, 4th edition, Pearson Education, 2018.

5.26 Proving the set of integers under subtraction is not a group

Let's begin with the definition of a group. A group is a set having the properties of closure and associativity under an operation; having an identity element; and having an inverse element for every member of the set. For further discussion of the definition of a group see section titled "Proving the uniqueness of inverse elements of groups."

An example of a group is the set of integers under addition. Whenever we add two integers together, we get another integer. Associativity is a feature of the group. The group identity element is zero. Every element of the group can be matched with its negative to yield the identity element when added together.

We are tempted to assume that if additivity over the integers is a group, then certainly subtraction over the integers must also be a group. But NO! Subtraction over the integers violates one of the defining properties of groups.

Let's consider the operation of subtraction, asking ourselves whether the group definition is fulfilled. Immediately, we find that subtraction lacks the associative property.

$$a - (b - c) \neq (a - b) - c$$

Because,

$$a - (b - c) = (a - b) + c$$

5.27 Proving the set of rational numbers under multiplication is not a group

1. In a group, every element of the group must have an inverse.
2. In the case of the rational numbers under multiplication, the identity element is 1.
3. Therefore, every element a of the group has an inverse element that we'll call a^{-1}that, when multiplied together, yields 1.
4. Zero is a rational number (i.e., it can be represented as a fraction of integers (e.g., $0/5 = 0$).
5. But the rational number zero has no inverse element (any rational number times 0 will always equal 0, and will never equal 1)
6. Therefore, **not every** rational element under multiplication has an inverse.
7. Therefore the set of rational numbers under multiplication is not a group.

We should note that the set of rational numbers with the element 0 removed **is a group** under the operation multiplication:

5.28 Proving the set of irrational numbers under multiplication is not a group

We've shown that the rational numbers under multiplication is not a group. Will we have any better luck when we try to determine whether the irrational numbers under multiplication is a group?

In section titled "Proving irrational numbers exist", we proved that $\sqrt{2}$ is an irrational number. Knowing this, we can easily show that the irrational numbers under multiplication is not a group:

1. We'll begin by assuming that the set of irrational numbers is a group.
2. We'll multiply $\sqrt{2}$ with itself to get 2.
3. But 2 is not irrational.
4. Therefore the group operation (multiplication) performed on an element of the group (of irrational numbers) violates closure.

   By definition of a group, the group operation performed on elements of the group must always yield another element of the same group.
5. Therefore, our original premise is false, and the irrational numbers are not a group under the operation of multiplication.

In this section and the preceding two sections, we have learned an important lesson. In group theory, it is the group operation that determines the properties of the group. It is sometimes possible to find two groups, with completely different sets of elements, but with group operations that provide a one-to-one mapping between the dissimilar elements of the two groups. When the mapping between the elements of the two groups are preserved after any and all group operations, the correspondence is known as an **isomorphism**. Isomorphisms play an important role in group theory insofar as the rules and theorems developed for a group will apply to any other group to which it is isomorphic. Some of the most advanced proofs in mathematics hinge upon demonstrating an isomorphism between one group and another; then applying the previously proven properties of the isomorphic group back to the group of interest. A good deal of modern physics involves establishing an isomorphism between a geometric group and a matrix group, then applying the well-studied techniques of matrix algebra to those elements in the matrix group that represent elements from the geometric group.

5.29 Proving the Fourier inversion theorem

Here is the famous Fourier transform (for continuous functions):

$$\hat{f}(\xi) = \int_{-\infty}^{\infty} f(x)e^{-2\pi i x\xi}\,dx \tag{5.1}$$

The Fourier transformation is a transformation that takes a continuous function and creates from it a sum of sinusoids. As such, it is simply a mathematical operation and does not require proof.

The utility of the Fourier transformation is found when we perform a transformation on the result of the Fourier transformation and retrieve our original function. That is to say, the inverse of the fourier transformation of a function is the function itself. Hence, it is the so-called Fourier inversion theorem that we need to prove (not the Fourier transform)

Symbolically, the Fourier inversion theorem is represented as:

$$\mathcal{F}^{-1}(\mathcal{F}f)(x) = f(x)$$

Here $\mathcal{F}$ is the Fourier transform and $\mathcal{F}^{-1}$ is its inverse.[8]

Substituting for $\mathcal{F}^{-1}(\mathcal{F}f)(x)$ our equations for the Fourier transform and its inverse, on a function, f(x), we have:

$$\begin{aligned}\mathcal{F}^{-1}(\mathcal{F}f)(x) &= \frac{1}{2\pi}\int_{-\infty}^{\infty} e^{ipx}\left(\int_{-\infty}^{\infty} e^{-ip\alpha} f(\alpha)\,d\alpha\right)dp \\ &= \frac{1}{2\pi}\int_{-\infty}^{\infty}\left(\int_{-\infty}^{\infty} e^{ipx}e^{-ip\alpha}\,dp\right)f(\alpha)\,d\alpha\end{aligned}$$

But the Dirac delta function can be expressed as: [9]

$$\delta(x-\alpha) = \frac{1}{2\pi}\int_{-\infty}^{\infty} e^{ip(x-\alpha)}\,dp$$

[8]Notice that we express the Fourier transform on f(x) as $(\mathcal{F}f)(x)$ and not $\mathcal{F}(f(x))$. The reason for this is that the Fourier transform operates on functions, not on variables. It is commonly employed to transform a function having a time variable into another function having a frequency variable.

[9]Here, we use the approach that the direct delta function is basically the squeezed Gaussian distribution, with its standard deviation approaching zero. See sections titled "Proving that $\int_{-\infty}^{\infty}\delta(x)dx = 1$", and "Proving Fourier transforms of Gaussians are Gaussians".

So:

$$\mathcal{F}^{-1}(\mathcal{F}f)(x) = \int_{-\infty}^{\infty} \delta(x-\alpha) f(\alpha)\, d\alpha$$

But, by the sifting property,[10]

$$f(x) = \int_{-\infty}^{\infty} \delta(x-\alpha) f(\alpha)\, d\alpha$$

Thus,

$$\mathcal{F}^{-1}(\mathcal{F}f)(x) = f(x)$$

And the inversion property of the Fourier transform is proven.

[10]See section titled "Proving the sifting property of Dirac delta".

5.30 Proving even and odd decomposition applies to all functions

We briefly used even and odd functions in the section titled "Proving the Pauli exclusion principle." Decomposing functions into even and odd components is, surprisingly, one of the most useful techniques in wave mechanics and in signal processing. We'll prove some of the fundamental properties of even and odd functions. Along the way, we'll catch a glimpse of how these properties can be applied.

A function is even if $f(x) = f(-x)$.

A function is odd if $f(x) = -f(-x)$.[11]

An example of an even function is $\cos(x)$. An example of an odd function is $\sin(x)$. We shall soon see that the Fourier series transforms functions into a series of even functions (cosines) and odd functions (sines). Thus, the Fourier series will (eventually) allow us to put into practice the proven properties of even and odd functions.

Let's prove that any function can be decomposed into even and odd functions.

First, by playing around with expressions, we can produce the following obvious relationship (which applies to any function):

$$f(x) = 1/2\Big(f(x) + f(-x) - f(-x) + f(x)\Big)$$

$$f(x) = 1/2\Big(f(x) + f(-x)\Big) + 1/2\Big(f(x) - f(-x)\Big)$$

Looking at the right side of the equation, we shall prove that $(f(x) + f(-x))$ is always even, and the second function term on the right side, $(f(x) - f(-x))$ is always odd. Once we've proven these two assertions, we will have shown that any function can be represented by $1/2$ of an even function plus $1/2$ of an odd function. Thus, any function can be decomposed into even and odd functions.

Let's look at $(f(x) + f(-x))$, which we'll call f_e

$$f_e(x) = (f(x) + f(-x))$$

[11]You may find that the terms "even" and "odd" functions appear, equivalently, as "symmetric" and "anti-symmetric" functions.

Let's substitute $-x$ for x.

$$f_e(-x) = (f(-x) + f(x)) = (f(x) + f(-x)) = f_e(x)$$

So $f_e(x) = f_e(-x)$ and, by definition, $f_e(x)$ is an even function.

Now, let's let f_o represent the second function term on the right side of the equation, $f_o(x) = (f(x) - f(-x))$. We have:

$$f_o(-x) = (f(x) - f(-x))$$

Let's substitute $-x$ for x.

$$f_o(-x) = (f(-x) - f(x)) = -(f(x) - f(-x)) = -f_o(x)$$

So $f_o(-x) = -f_o(x)$ and, by definition, $f_o(x)$ is an odd function.

Therefore, any function can be expressed as the sum of even and odd functions.

5.31 Proving the integral of even times odd parts of a function is zero

Using the notation adopted for the prior section, we need to prove that:

$$\int_{-\infty}^{\infty} f_e(x) f_o(x) dx = 0$$

Let's begin by evaluating the integral over two intervals, from negative infinity to zero and from zero to infinity.

$$\int_{-\infty}^{\infty} f_e(x) f_o(x) dx = \int_{-\infty}^{0} f_e(x) f_o(x) dx + \int_{0}^{\infty} f_e(x) f_o(x) dx$$

In the first integral on the right side of the equation, we'll substitute -x for x (and dx becomes -dx). This substitution does not change the overall value of the equation, but it does yield the negative value of the odd function, the negative value of the integral, reverses the interval of integration and changes the sign of the interval of integration (four negatives that all cancel out to a positive).

Leaving us with,

$$\int_{-\infty}^{\infty} f_e(x) f_o(x) dx = \int_{0}^{\infty} f_e(-x) f_o(-x) dx + \int_{0}^{\infty} f_e(x) f_o(x) dx$$

Rearranging,

$$\int_{-\infty}^{\infty} f_e(x) f_o(x) dx = \int_{0}^{\infty} \Big(f_e(-x) f_o(-x) + f_e(x) f_o(x) \Big) dx$$

By definitions of even and odd functions, we can safely substitute $f_e(-x)$ for $f_e(x)$ and $-f_o(-x)$ for $f_o(x)$, yielding:

$$\begin{aligned} \int_{-\infty}^{\infty} f_e(x) f_o(x) dx \quad &= \int_{0}^{\infty} \Big(f_e(-x) f_o(-x) - f_e(-x) f_o(-x) \Big) dx \\ &= \int_{0}^{\infty} 0 dx \\ &= 0 \end{aligned}$$

This proves the integral of even times odd components of a function is zero.

5.32 Proving the Fourier even/odd split

We have previously shown that functions can always be decomposed into even and odd functions (see section titled "Proving even and odd decomposition applies to all functions".) We must take note that a transform is NOT a function. A transform is an operation on a function, and we cannot assume that transforms can be decomposed into the sum of an even and an odd transform. We need to prove it so.

In this section, we will show that any Fourier transform can be expressed as the sum of an even transform and an odd transform. To do so, we'll rely upon two proofs that we have encountered in prior sections. To start, let's review a bit about Fourier transforms.

The Fourier transform on a function, $(f(t))$, is:

$$F(s) = \int_{-\infty}^{\infty} f(t)e^{-i2\pi st}dt$$

From Euler's formula, and from our knowledge that cosine is an even function and sine is an odd function,

$$e^{-i2\pi st} = cos(-2\pi st) + isin(-2\pi st) = cos(2\pi st) - isin(2\pi st)$$

So,

$$F(s) = \int_{-\infty}^{\infty} f(t)cos(2\pi st)dt - i\int_{-\infty}^{\infty} f(t)sin(2\pi st)dt$$

But we know that f(t) can be decomposed into the sum of an even and an odd function (which we shall call $f_e(t)$ and $f_o(t)$, respectively). We shall let $f(t) = f_e(t) + f_o(t)$ and substitute back into our equation:

$$\begin{aligned} F(s) = &\int_{-\infty}^{\infty} f_e(t)cos(2\pi st)dt \\ &+ \int_{-\infty}^{\infty} f_o(t)cos(2\pi st)dt \\ &-i\int_{-\infty}^{\infty} f_e(t)sin(2\pi st)dt \\ &-i\int_{-\infty}^{\infty} f_o(t)sin(2\pi st)dt \end{aligned}$$

We know that the integral of an even function multiplied by an odd function is always zero.[12] We also know that cosine is an even function and sine is an odd function, so the middle components inside two integrals evaluate to zero.

Hence,

$$F(s) = \int_{-\infty}^{\infty} f_e(t)cos(2\pi st)dt - i\int_{-\infty}^{\infty} f_o(t)sin(2\pi st)dt$$

Or,

$$F(s) = F_e(s) + F_o(s)$$

Therefore, a Fourier transform can decompose into an even and odd Fourier transform.

This outcome may appear to be trivial, but the process of even and odd decomposition can be carried through iteratively, and this computational opportunity plays a key role in the calculation of the Fast Fourier Transform (FFT). Moreover, even/odd transform decomposition is a stalwart technique employed in digital signal processing and analysis.

[12]See section titled "Proving the integral of an even function times an odd function is always zero".

5.33 Proving the Goldbach Conjecture, partially

The Goldbach conjecture is an assertion raised in 1742 by the Russian mathematician Christian Goldbach (1690-1764). The conjecture states that every even integer larger than 2 can be expressed as the sum of two prime numbers. For example 10 = 7 + 3 and 18 = 11 + 7. To date, nobody has found an even integer that violates the conjecture, but nobody has actually managed to prove that the conjecture is always true. The following argument falls short of "proof", but it manages to show that the conjecture is true for an infinite number of even integers.

1. We know that there are even integers (x) such that $x/2 + 1 = p$, where p is a prime number.

   All prime numbers greater than 2 are expressible as an integer plus 1. We know that if x is an even integer, then $x/2$ must be an integer (that is what it means to be an even integer). Therefore p must be an integer, and we have every right to specify that we choose our value of x such that $x/2 + 1$ is some prime number, p.

2. We place one more constraint on x, requiring that $x/2 - 1 = q$ where q is another prime number.

   The same reasoning applies as in step 1, only we'll be subtracting 1 from $x/2$

3. Now let's add the first assertion to the second.

   $x/2 + 1 = p$

   plus

   $x/2 - 1 = q$

   gives us

   $x = p + q$

   Therefore,

4. When x is an even integer, we can be certain that x can be expressed as the sum of two prime numbers. But we can choose an infinite number of values of x (we have an infinite number of even integers and primes from which to choose)

Therefore,

5. The Goldbach conjecture is true for an infinite number of integers

Sadly, we come up short of providing the proof that **every** even integer can be expressed as the sum of two prime integers. But it's a start.

5.34 Proving the Gaussian integral $= \sqrt{\pi}$

The Gaussian integral, named for Carl Friedrich Gauss (1777-1855) represents the so-called normal probability distribution, producing the familiar bell-shaped curve.

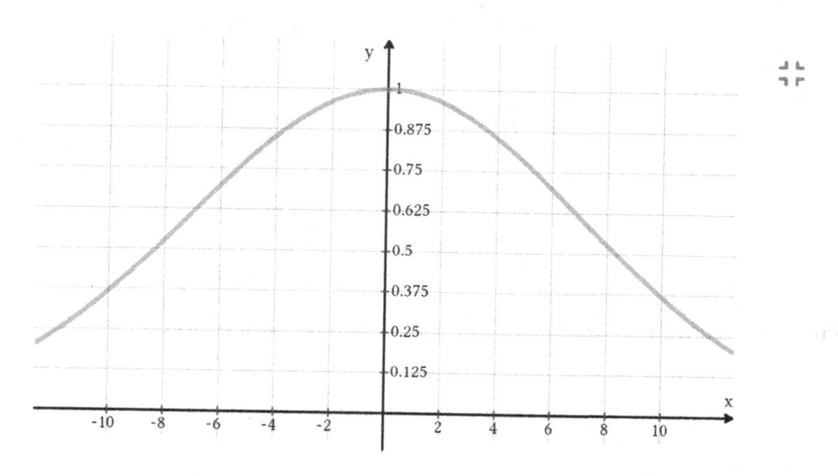

FIGURE 5.8: The familiar bell-shaped curve of the Gaussian distribution.

$$\int e^{-x^2}\, dx \qquad \text{Gaussian integral}$$

We can easily prove that the area under the Gaussian, evaluated from $-\infty$ to ∞ is $\sqrt{\pi}$, shown here as an equation:

$$\int_{-\infty}^{\infty} e^{-x^2}\, dx = \sqrt{\pi}$$

Let's square the Gaussian integral and convert it to radial coordinates:

$$
\begin{aligned}
\iint_{\mathbb{R}^2} e^{-(x^2+y^2)}\,dx\,dy &= \int_0^{2\pi}\int_0^{\infty} e^{-r^2}\,r\,dr\,d\theta \\
&= 2\pi \int_0^{\infty} re^{-r^2}\,dr \\
&= 2\pi \int_{-\infty}^{0} \tfrac{1}{2}e^{s}\,ds \qquad \text{substituting } s = -r^2 \\
&= \pi \int_{-\infty}^{0} e^{s}\,ds \\
&= \pi\left(e^{0} - e^{-\infty}\right) \\
&= \pi,
\end{aligned}
$$

Taking the square root of both sides of the equation, we have:

$$\int_{-\infty}^{\infty} e^{-x^2}\,dx = \sqrt{\pi}$$

Having shown that the Gaussian integral $= \sqrt{\pi}$, we can derive another very handy formula that will be used extensively to solve various wave equations; namely, that $\int_{-\infty}^{\infty} e^{-\pi t^2}dt = 1$.

We start with:

$$\int_{-\infty}^{\infty} e^{-x^2}\,dx = \sqrt{\pi}$$

We let $x = \sqrt{\pi}t$ and $dx = \sqrt{\pi}dt$. Substituting back into the Gaussian integral:

$$\sqrt{\pi}\int_{-\infty}^{\infty} e^{-\pi t^2}dt = \sqrt{\pi}$$

Then,

$$\int_{-\infty}^{\infty} e^{-\pi t^2}dt = 1$$

5.35 Proving the divergence theorem, informally

The divergence theorem, also known as Gauss's theorem or Ostrogradsky's theorem equates the flux of a vector field through a closed surface with the divergence of the vector field within the enclosed volume. An intuitive and logical argument for the theorem (which is all that we will be developing in this section) can be summarized in a few sentences, but to get any insight into the theorem, we need to quickly review the meaning of a vector field and the meaning of the divergence of a vector field.

A vector field is an assignment of a vector (something with a value and a direction) to each point in a defined space (i.e., the vector field).[13] In three dimensional space, each point can be assigned a value for each of the three spatial dimensions. A vector at a point in 3-D Euclidean space can be represented by:

$$\vec{f} = f_1\vec{i} + f_2\vec{j} + f_3\vec{k}$$

Where $\vec{f}$ is a vector and $\vec{i}$, $\vec{j}$, and $\vec{k}$ are unit directional vectors for the x,y, and z coordinates. The divergence of a vector field is:

$$\operatorname{div}\mathbf{F} = \nabla \cdot \mathbf{F} = \frac{\partial F_1}{\partial x} + \frac{\partial F_2}{\partial y} + \frac{\partial F_3}{\partial z},$$

At any given point in the field, the divergence at the point represents the flow from the point (i.e. a measurement of the source, positive or negative, at the point). The volume integral of the sources tells us how much net flow exits (or enters) the enclosed volume.

The sum of all of the sources (of the measured vector) at every point within an enclosed field must equal the net flux through the enclosure. In simple words, whatever exits the enclosure must have come from what's inside, and what's inside is represented by the vectors at every point in the enclosed field.

The divergence theorem formula is:

$$\iiint\limits_V (\nabla \cdot \mathbf{F})\, \mathrm{d}V = \iint\limits_S (\mathbf{F} \cdot \hat{\mathbf{n}})\, \mathrm{d}S.$$

The left-hand side of the equation is the integrated sum of the field vectors

[13]When we discuss fields of forces (such as the electromagnetic field or the gravity field), we are referring to vector fields in unbounded spacetime.

in the enclosed volume. The right-hand side of the equation is the summed fluxes through the enclosure. Although there is a robust mathematical proof for the divergence theorem, we will dispense with the formalities, accepting that whatever leaves an enclosed vector field must have originated inside the vector field. The divergence theorem permits us to calculate flux out of a surface by taking the divergence of the vector field integrating over its volume (in xyz coordinates, the volume increment, dV, becomes dxdydz)

We note that the general continuity equation of physics restates the divergence theorem of calculus. The continuity equation expresses the relationship between conserved quantities moving within a boundary and the net flux out of the boundary. It appears in the following form for a boundary over which the net flux out is zero (i.e., when there is a conserved quantity within the boundary).

$$\frac{\partial \rho}{\partial t} = -\nabla \cdot \mathbf{J} \qquad \text{continuity equation for a conserved quantity}$$

When the quantity is not conserved, then the continuity equation would appear as:

$$\frac{\partial \rho}{\partial t} + \nabla \cdot \mathbf{J} = \sigma \qquad \text{General continuity equation}$$

The symbols of the equation translate to:

- ρ is the density of the conserved quantity within the boundary (the quantity per volume inside the boundary)
- j is the flux density of the conserved quantity (the amount of charge passing through a unit area of the boundary)
- σ is the generation of the quantity per unit volume per unit time
- t is time
- $\nabla \cdot$ is the divergence

The continuity equation applies under the condition of a closed boundary, but how might we apply the continuity equation for all of space? No problem. We just enlarge the boundary to fill the universe. In this case, flux vanishes at infinity. If the flux vanishes, then we can expect no change in the total system. So, for example, in the case of charge, we would infer that the total charge of the infinite system must be conserved.

5.36 Proving the volume of revolution formula

The volume of revolution formula states that the volume of a solid formed by rotating the portion of a curve (between points and b) around the x axis is $V = \int_a^b \pi y^2 dx$

The proof is trivial once we can visualize the volume of the solid as a stack of disks, with each disk having a disk radius $y = f(x)$ and a disk volume of its area (approximating a circle) times its height (Δx), or $\pi y^2 \Delta x$.

So, the volume is just the sum of the volume of all the disks.

$$V = \sum_a^b \pi y^2 \Delta x$$

When we allow δx to approach a limit of zero, we can represent the volume as an integral.

$$V = \int_a^b \pi y^2 dx$$

This is the general formula for the volume of rotation.

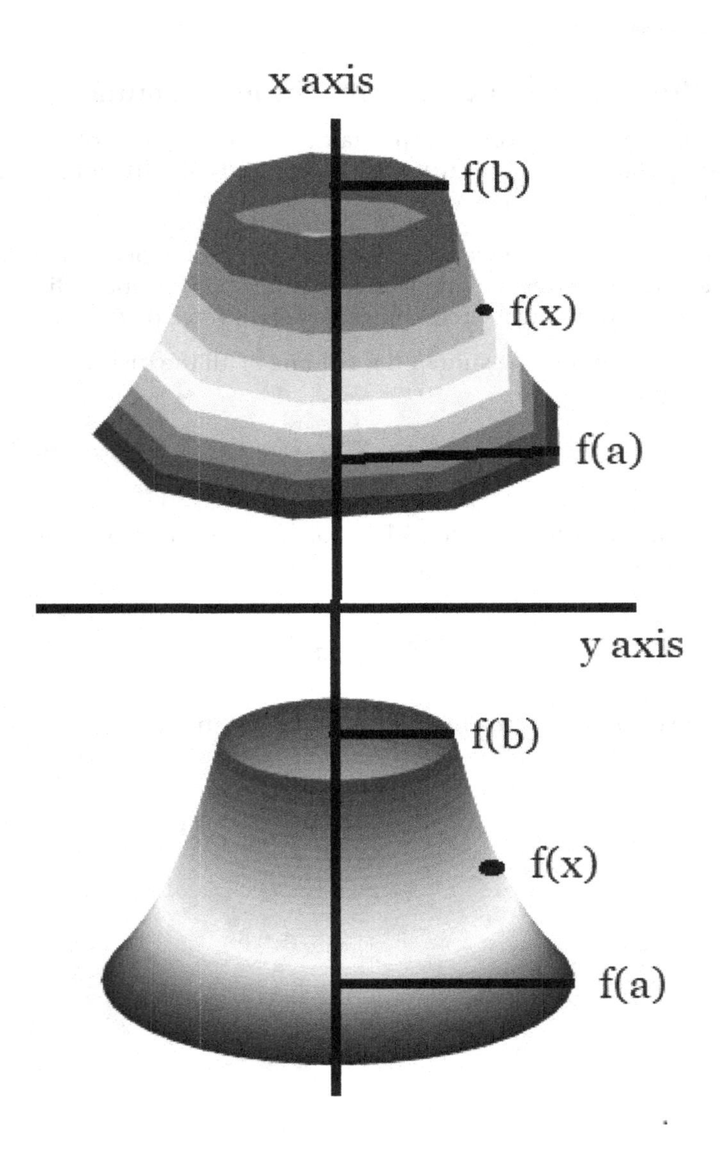

FIGURE 5.9: A volume is created by revolving a curve about a central axis. At the top of the figure, a volume is divided into 10 discrete disks, each of whose volume is approximated by the area of the base of the disk times the height of the disk. The total volume is the sum of the volumes of all ten disks. At the bottom of the figure, the volume is equivalent to the limit achieved when the large number of disks of negligible height are summed together.

5.37 Proving the volume of a sphere is $4/3\pi r^3$

We'll begin by creating a sphere by rotating a semicircle, centered at x = 0, about the x-axis, as shown in our figure.

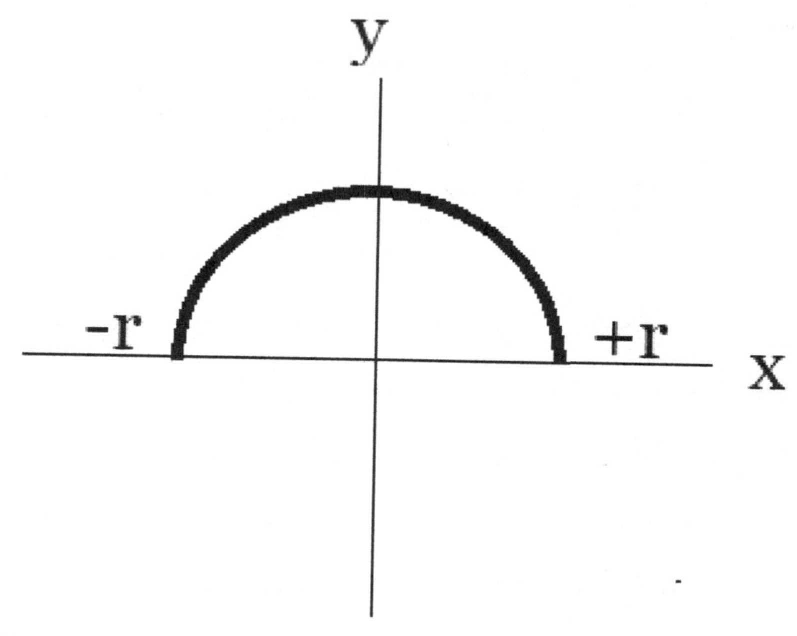

FIGURE 5.10: A sphere is produced when we rotate a semicircle, with a radius r, about the x-axis.

We can use our formula for calculating a volume of revolution proven in the prior section (i.e., the volume of revolution formula states that the volume of a solid formed by rotating the portion of a curve around the x-axis is $V = \int_a^b \pi y^2 dx$). In this case, we have:

$$V = \int_{-r}^{r} \pi y^2 dx$$

For a circle, we know that $x^2 + y^2 = r^2$, and that r, the radius, is a constant value.

Rearranging $y^2 = r^2 - x^2$.

$$V = \int_{-r}^{r} \pi(r^2 - x^2)dx$$

or

$$V = \int_{-r}^{r} \pi r^2 dx - \int_{-r}^{r} \pi x^2 dx$$

Evaluating:

$$V = 2\pi r^3 - \left.\frac{\pi r^3 x}{3}\right|_{-r}^{r} = 2\pi r^3 - \frac{2\pi r^3}{3} = 4/3\pi r^3$$

5.38 Proving Bayes' theorem

When two events, A and B occur independently of one another, the probability of both occurring is just the probability of A occurring times the probability of B occurring.

When the occurrence of A and B occurs dependently of one another, how do we calculate their occurrences? Thomas Bayes gave us a simple theorem that tells us how to calculate dependent probabilities.

The probability of two dependent events A and B occurring is represented $P(A\&B)$. It is equal to the probability of A occurring times the probability of B occurring on the condition that A has occurred (the latter being represented as $P(B|A)$. As an equation, it is shown as,

$$P(A\&B) = P(A) \cdot P(B|A)$$

We also know that the same combined probability must be equal to the probability of B occurring times the probability that A occurs on the condition that B has occurred (because we're finding the probablity that both A and B occur).

$$P(A\&B) = P(B) \cdot P(A|B)$$

Equating the two gives us,

$$P(B) \cdot P(A|B) = P(A) \cdot P(B|A)$$

Dividing both sides by P(B) yields,

$$P(A|B) = \frac{P(A) \cdot P(B|A)}{P(B)} \qquad \text{Bayes theorem}$$

5.39 Proving the Taylor expansion

Here is the Taylor expansion (also known as the Taylor approximation and the Taylor series, and named for Brook Taylor, 1685–1731).

The Taylor expansion

$$f(x) = f(x_0) + f'(x_0)(x - x_0) +$$
$$\frac{f''(x_0)}{2!}(x - x_0)^2 +$$
$$\frac{f'''(x_0)}{3!}(x - x_0)^3 + \cdots,$$

In compact form,

$$\sum_{n=0}^{\infty} \frac{f^{(n)}(a)}{n!}(x - a)^n,$$

For functions that are easy to differentiate, but difficult to directly evaluate, the Taylor expansion is a handy tool. Amazingly, its derivation only requires the fundamental theorem of calculus and repeated application of integration by parts (shown below).

$$\int_a^x f'(t) = f(x) - f(a)$$

When we set $a = 0$, we have,

$$\int_0^x f'(t)dt = f(x) - f(0) \qquad \text{Fundamental theorem of calculus}$$

Or,

$$f(x) = f(0) + \int_0^x f'(t)dt$$

The equation for integration by parts is,

$$\int u dv = uv - \int v du$$

Alternately expressed,

$$\int v'u = vu|_a^b + \int vu' \qquad \text{Integration by parts} \tag{5.2}$$

Let's apply them to a differentiable function f(x) and see what we get. Let's look once more at the Rearranging the fundamental theorem of calculus,

$$f(x) = f(0) + \int_0^x f'(t)dt$$

Now, we'll take the integral and use the formula for integration by parts, with $u = f'(t)$ and $v' = dt$, to yield,

$$f(x) = f(0) + t|_0^x f'(0) \int_0^x (t/1)f''(t)dt$$

Evaluating,

$$f(x) = f(0) + (x/1)f'(0) + \int_0^x tf''(t)dt$$

We can apply integration by parts again, to the integral on the right side of the equation, using $v' = tdt, v = t^2/2$, $u = f''$ and $du = f'''$.

$$f(x) = f(0) + (x/1)f'(0) + (x^2/2 \cdot 1)f''(0) + \int_0^x (t^2/2 \cdot 1) \cdot f'''dt$$

Applying integration by parts once more,

$$f(x) = f(0) + xf'(0) + (x^2/2 \cdot 1)f''(0) + (x^3/3 \cdot 2 \cdot 1)f''' + \int_0^x (t^3/3 \cdot 2 \cdot 1)f''''dt$$

Repeating this step indefinitely yields the complete Taylor expansion. We can use the Taylor expansion to evaluate any differentiable function. The Taylor expansion of the exponential function is,

$$e^x = \sum_{n=0}^{\infty} \frac{x^n}{n!} = 1 + x + \frac{x^2}{2!} + \frac{x^3}{3!} + \cdots$$

The Taylor expansion of the sine wave (remembering that f(0) evaluates to zero when its sine is at zero degrees, and that the derivative of a sine is its cosine, and the derivative of a cosine is its negative sine.) is,

$$\sin(x) \approx x - \frac{x^3}{3!} + \frac{x^5}{5!} - \frac{x^7}{7!}.$$

We can demonstrate that we do not need to compute many of the terms of the Taylor expansion to get a fairly good approximation of the sin(x) function. Two or three terms will suffice. Here are the Gnuplot command lines that compute three approximations for the sin(x) function, using just the first two terms in the expansion, or the first three terms, or the first four terms. The lines of Gnuplot code that establish the embellishments of the output graph are omitted.

```
approx_1(x) = x - x**3/6
approx_2(x) = x - x**3/6 + x**5/120
approx_3(x) = x - x**3/6 + x**5/120 - x**7/5040
GPFUN_approx_1="approx_1(x) = x - x**3/6"
GPFUN_approx_2="approx_2(x) = x - x**3/6 + x**5/120"
GPFUN_approx_3="approx_3(x) = x - x**3/6 + x**5/120 - x**7/5040"
label1 = ‘‘x - {x^3}/3!"
label2 = ‘‘x - {x^3}/3! + {x^5}/5!"
label3 = ‘‘x - {x^3}/3! + {x^5}/5! - {x^7}/7!"
plot ’+’ using 1:(sin($1)):(approx_1($1)) with filledcurve title
label1 lt rgb ‘‘blue’’ ,      ’+’ using 1:(sin($1)):(approx_2($1))
with filledcurve title label2 lt rgb ‘‘green’’,      ’+’ using
1:(sin($1)):(approx_3($1)) with filledcurve title label3 lt rgb
"red’’,      sin(x) with lines lw 1 lc rgb ‘‘black"
```

The first few components of a Taylor expansion often serve as a fairly good approximation of the infinite sum,

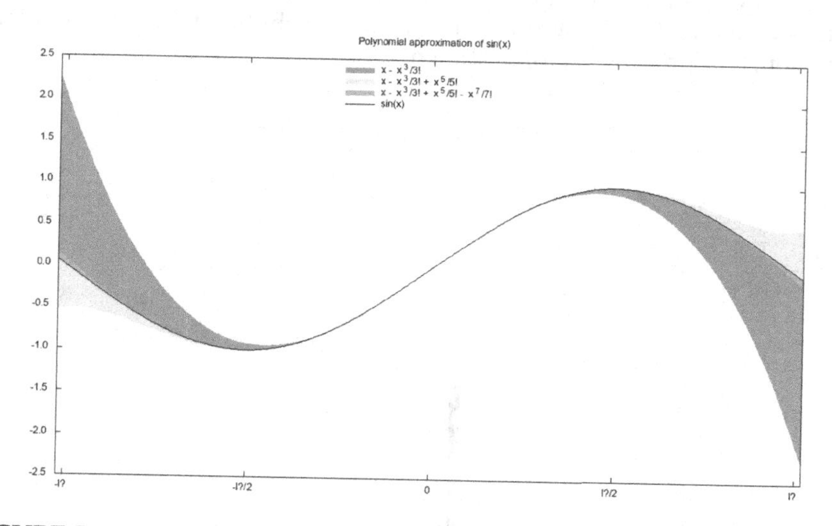

FIGURE 5.11: Approximation of a sine wave using 2, 3, or 4 of the first terms of the Taylor expansion of the sine wave function, an infinite sum of polynomials. As we incorporate more terms into the approximation, we come closer to a true sin(x) function. We need only use the first few terms of the Taylor expansion to produce a fairly faithful sine wave.

5.40 Proving $\int_{-\infty}^{\infty} \delta(x)dx = 1$

$\delta(x)$ is the Dirac distribution (sometimes referred to as the Dirac delta or, less accurately, as the Dirac delta function). It is a distribution whose value for x is 0 everywhere along the x-axis **except for one point** where the value of x is 1. A fairly good way of thinking of the Dirac distribution is that it is the thinnest possible normal distribution. Imagine a normal distribution that is squeezed in from its left and right sides so tightly that it consists of a straight line at one point, and zero everywhere else. In other words, the Dirac delta is equivalent to the normal (statistical) distribution when the standard deviation, σ approaches zero, as shown:

$$\delta(x-\mu) = lim_{\sigma \to 0} \frac{1}{\sqrt{2\pi\sigma^2}} e^{\frac{-(x-\mu)^2}{2\sigma^2}}$$

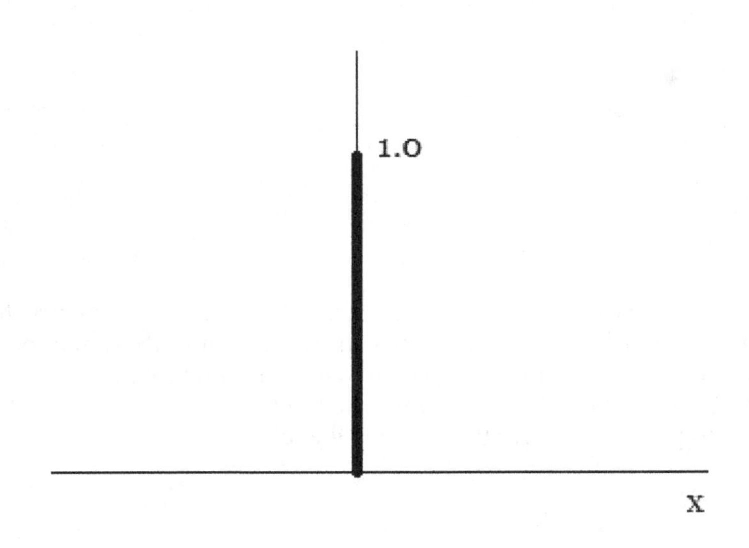

FIGURE 5.12: Dirac delta. The value of $\delta(x)$ is 1 where x = 0. Elsewhere, the value of $\delta(x)$ is zero.

We can use the equivalence between the extreme normal distribution and the Dirac delta to prove that the integral of the Dirac delta distribution must equal 1.

The general equation for a normal distribution is:

$$f(x) = \frac{1}{\sigma\sqrt{2\pi}} e^{\frac{-(x-\mu)^2}{2\sigma^2}}$$

Let's evaluate f(x) using parameters that μ, setting μ, the mean, at the center (zero coordinate) of the x-axis.

$$f(x) = \frac{1}{\sigma\sqrt{2\pi}} e^{\frac{-(x)^2}{2\sigma^2}}$$

Let's evaluate the integral of the normal distribution:

$$\int_{-\infty}^{\infty} \frac{1}{\sigma\sqrt{2\pi}} e^{\frac{-(x)^2}{2\sigma^2}}$$

Let's substitute $y = x/(\sqrt{2}\sigma)$, so that $dy = dx/(\sqrt{2}\sigma)$ and $dx = \sqrt{2}\sigma dy$

$$\int_{-\infty}^{\infty} \frac{1}{\sigma\sqrt{2\pi}} e^{-(y)^2} \sqrt{2}\sigma dy$$

This evaluates to:

$$\frac{1}{\sqrt{\pi}} \int_{-\infty}^{\infty} e^{-(y)^2} dy$$

But we have previously shown, in Section titled "Proving that the Gaussian integral = $\sqrt{\pi}$" that $\int_{-\infty}^{\infty} e^{-(y)^2} dy = \sqrt{\pi}$.

Therefore,

$$\frac{1}{\sqrt{\pi}} \int_{-\infty}^{\infty} e^{-(y)^2} dy = \frac{\sqrt{\pi}}{\sqrt{\pi}} = 1$$

The area of the curve under the full normal distribution equals 1 for all normal distributions. We have stipulated that the Dirac delta distribution is an extreme version of the normal distribution wherein all possible values of x are squeezed into a single coordinate, the mean value. Therefore, the integral of the Direct delta function must also equal 1 and:

$$\int_{-\infty}^{\infty} \delta(x)dx = 1$$

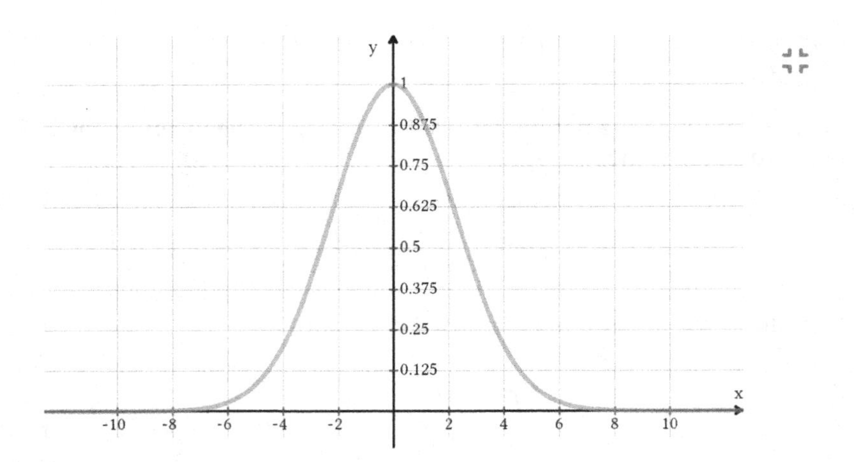

FIGURE 5.13: The typical bell-shaped curve of a normal distribution.

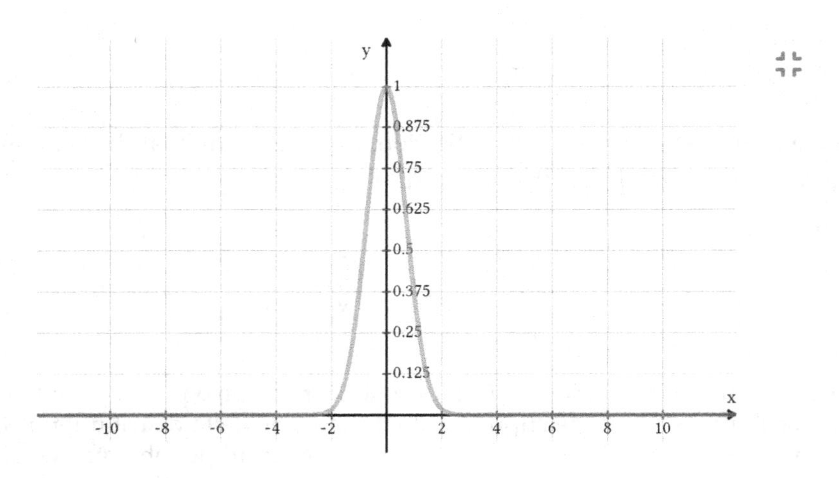

FIGURE 5.14: A squeezed normal distribution. As the standard deviation approaches zero, the normal distribution approaches the Dirac delta distribution.

5.41 Proving the sifting property of Dirac delta

The sifting property of the impulse states that when f(t) is integrated over all possible values, the impulse function included in the integral sifts the value of f(t) at the point when the time equals a (i.e., f(a)). In the form of a general equation:

$$\int_{-\infty}^{\infty} f(t)\delta(t-a)dt = f(a)$$

The proof is trivial and depends on knowing two items,

- $\int_{-\infty}^{\infty} \delta(t-a)dt = 1$

  See section titled "Proving $\int_{-\infty}^{\infty} \delta(x)dx = 1$",

- $\delta(t-a)$ is zero everywhere except at the point when $t = a$

  This is the defining property of the delta function.

Combining the two items,

$$\int_{-\infty}^{\infty} f(t)\delta(t-a)dt = f(a)\int_{-\infty}^{\infty} \delta(t-a)dt = f(a)$$

5.42 Proving the convolution theorem

The convolution theorem states that the Fourier transform of the convolution of two functions equals the product of the Fourier transforms of the functions.

In mathematical terms:

$$\mathcal{F}(f \star g) = \mathcal{F}(f)\mathcal{F}g = \hat{f} \cdot \hat{g}$$

Here, $\mathcal{F}$ is the Fourier transform and $\hat{f}$ or $\hat{g}$ is a shorthand way of expressing the Fourier transform of f or g. As noted previously, the Fourier transform takes a function in the time domain and yields a function in the frequency domain. The inverse Fourier transform does the reverse. Using the equation above, we can take the inverse Fourier transform of $\hat{f}\hat{g}$, to yield the convolution of f and g.

Consequently,

$$\mathcal{F}^{-1}(\hat{f} \cdot \hat{g}) = f \star g$$

The convolution theorem is often summarized by saying that convolution in one domain (time or frequency) is equivalent to multiplication in the other domain.

Before we begin our proof of the convolution theorem, let's review the general equations for a Fourier transform and its inverse, and the convolution of two functions.

The Fourier transform of a function:

$$\mathcal{F}(f(x)) = \int_{-\infty}^{\infty} f(x)e^{-i\omega x}dx = \hat{f}(\omega)$$

The inverse Fourier transform of the same function

$$\mathcal{F}^{-1}(\hat{f}(\omega)) = \frac{1}{2\pi}\int_{-\infty}^{\infty} \hat{f}(\omega)e^{i\omega x}d\omega = f(x)$$

The definition of a convolution,[14]

$$f \star g = \int_{-\infty}^{\infty} f(x-\xi)g(\xi)d\xi$$

[14]We won't discuss convolution other than to define it mathematically.

The three listed equations are all that we need for our proof.

$$\mathcal{F}^{-1}(\hat{f}\hat{g})$$
$$= \frac{1}{2\pi}\int_{-\infty}^{\infty} \hat{f}(\omega)\hat{g}(\omega)e^{i\omega x}d\omega$$
$$= \frac{1}{2\pi}\int_{-\infty}^{\infty} \hat{f}(\omega)\left(\int_{-\infty}^{\infty} g(y)e^{-i\omega y}dy\right)e^{i\omega x}d\omega$$

Evaluate the Fourier transform of $\hat{g}(\omega)$

$$= \int_{-\infty}^{\infty} g(y) \times \frac{1}{2\pi}\int_{-\infty}^{\infty} \hat{f}(\omega)e^{i\omega(x-y)}d\omega dy \tag{5.3}$$

But $\frac{1}{2\pi}\int_{-\infty}^{\infty} \hat{f}(\omega)e^{i\omega(x-y)}d\omega$ is the inverse of the Fourier transform of f(x-y), yielding back itself. So, we can substitute f(x-y) back into our last equation above:

$$= \int_{-\infty}^{\infty} g(y)f(x-y)dy$$
$$= (g \star f)$$
$$= (f \star g)$$

This completes our proof

The very last step in the proof assumes that convolution is commutative, which we prove in the section titled "Proving convolution commutes."

5.43 Proving the impulse response convolution theorem

For our proof, we must begin with some understanding of linear systems. The following six properties of linear systems apply to all linear systems: [15]

1. **Additivity.**

   When we can express the properties of a system in terms of some function, f(), and components x and y, then additivity tells us that $f(x) + f(y) = f(x + y)$

2. **Homogeneity.**

   When we scale a component by "a", we scale the function by the same factor (i.e, $f(ax) = af(x)$)

3. **Superpositioning.**

   We can add and scale ecomponents and expect the outcome to be the same as adding and scaling functions (i.e., $f(ax + by) = f(ax) + f(by) = af(x) + bf(y)$. In the case of wave functions, superpositioning is equivalent to saying that two waves can exist at the same location at the same time, insofar as combining the waves produces the same outcome as adding the two waves separately.

4. **Shift invariance.**

   If we have two different functions that represent properties of the system, f() and y(), and we know that $f(x) = y(x)$, then shift invariance applies if $f(x + n) = y(x + n)$. For systems that accept functions as input to produce other functions as output, we can think of y() as the output of f().

5. **Sinusoidal fidelity.**

   For systems that accept functions as input to produce other functions as output, sinusoidal fidelity guarantees that a sinusoidal input will produce a sinusoidal output of the same frequency as the input.

[15] Linear systems have foundational importance in the areas of signal processing, signal analysis, and modern communications. The same mathematical formulae developed for the analysis of electromagnetic waves have seamlessly transitioned to the digital world. Today, any signal that varies over time (i.e., a waveform) can be sampled and converted to digital information, which can analyzed, modified, stored, transported, and reconstructed as a waveform.

6. **Linear modeling.**

   We can find functions that, when applied to the system under study, fulfill the properties of additivity and homogeneity. The representation of waves as sums of sines and cosines is a linear model.

Waves are linear systems, and everyday observations confirm that light waves behave linearly:

- When we increase the intensity of a lamp, the effects of the emitted light is purely additive. The increased photons don't compete with one another for space in the emerging light beam.
- When we turn on two different lamps in our living room, the effect on the lighting is additive, despite the difference in their locations. The two sources of light do not diminish one another's effect.
- When multiple beams of light collide, each separate beam emerges from the fray unscathed, moving with its original intensity, along its original course. and at the original speed.

Waves have all the properties of linear systems. As per property 6 (linear modeling), we know that there are functions that, when applied to a linear system, yield a result that is also a linear system. Let us suppose that $\mathcal{H}$ is such a function. Hence, all of the properties of linear systems apply to the products of the function.

Linear system homogeneity accounts for the scaling rule:

$$\mathcal{H}[ax(t)] = a\mathcal{H}[x(t)]$$

Linear system additivity accounts for the additive rule:

$$\mathcal{H}[x_1(t) + x_2(t)] = \mathcal{H}[x_1(t)] + \mathcal{H}[x_2(t)]$$

Together, the scaling rule and the additive rule justify the superpositioning rule, which we'll use in our proof: [16]

$$\mathcal{H}[ax_1(t) + bx_2(t)] = a\mathcal{H}[x_1(t)] + b\mathcal{H}[x_2(t)]$$

We add that time-shift invariance is a special form of shift invariance and accounts for the following:

$$\mathcal{H}[x(t - t_0)] = y(t - t_0) \qquad \text{Here, y() is the result of} \mathcal{H}[x()]$$

[16]Of the six properties of linear systems, the most important is "wave superpositioning"), insofar as any system for which superpositioning applies will always fulfill the minimal requirements for a linear system (i.e., additivity and homogeneity), and will nearly always fulfill all of the remaining properties of a linear system.

There is one more equality to remember. From the sifting property (which we have derived previously in the section titled "Proving the sifting property of Dirac delta"),

$$x(t) = \int_{-\infty}^{\infty} x(\tau)\delta(t-\tau)d\tau$$

Here, δ is the Dirac impulse function.

Now that we've dispensed with the preliminaries, we can easily prove the impulse response convolution theorem.

$$\begin{aligned}
y(t) &= \mathcal{H}[x(t)] && \text{Here, y(t) is the } \mathcal{H} \text{ operation on x(t)} \\
&= \mathcal{H}\left[\int_{-\infty}^{\infty} x(\tau)\delta(t-\tau)d\tau\right] && \text{Substitute the sifting identity for x(t)} \\
&= \int_{-\infty}^{\infty} \mathcal{H}\left[x(\tau)\delta(t-\tau)d\tau\right] && \text{Linear additivity property} \\
&= \int_{-\infty}^{\infty} x(\tau)\mathcal{H}[\delta(t-\tau)]d\tau && \text{Scaling property} \\
&= \int_{-\infty}^{\infty} x(\tau)h(t-\tau)d\tau && \text{Where h(t) is } \mathcal{H}[\delta(t)] \\
&= (x \star h)(t) && \text{The convolution of x with} \mathcal{H}[\delta]
\end{aligned}$$

Using just the fundamental properties of linear systems, and the sifting property, we have shown that to find a transformation of any component of the linear system (or any combination of components of the linear system), all we need to do is to convolve the component with the transformation of the Dirac delta function. This is the the impulse response convolution theorem, and, along with Fourier analysis (to which it is closely related), forms the core of all modern signal analysis. It appears frequently whenever linear systems are analyzed, and this would include all waves and harmonic oscillators of the Standard Model and of gravity.

5.44 Proving the principle of mathematical induction

The principle of mathematical induction holds that assertions regarding the set of positive integers, N, are true for every member, n, of the set N, if the following two statements are true:

1. The assertion is true for the case when n = 1.
2. For every positive integer k, when the assertion is true for n = k, the assertion is also true for n = k +1

The principle of mathematical induction would be proven wrong if there were any instances (of positive integers) wherein the above two conditions were true, but for which the assertion would not hold true.

- Let's assume we found a non-empty set, F, of positive integers for which the principle of mathematical induction were **false**.

  That is, when the principle holds true for n=1 and n=k but does not hold true for n=k+1, in F.

- The set F must have some positive integer that is the smallest integer belonging to the set.
- We know that whatever number this smallest number is, it cannot be "1" because we began by stipulating that our principle is true for n = 1.
- Therefore, the smallest integer for which the principle is false must be at least as large as 2.

  We'll call "s" the smallest member of F.

- We can infer that the principle is true for the case when n = s - 1.

  This is true because we chose s to the the smallest integer in the set of positive integers for which the assertion is false. Therefore, the assertion must hold true for any integer smaller than s.

- But if the assertion is true for n = s - 1, then the assertion must be true for n = s.

  This is true because the principle is true for all integers **other** than those included in set F, and we know that s-1 is **not** included in set F. This means that for s-1, the principle must be true for n = (s-1)+1 = s.

- This contradicts our assumption that the principle is false for n = s.

  Therefore,

- There can be no exceptions among the positive integers whenever the two conditions listed are met.

  The two conditions being are that when the principle is true for the case when n=1, and when n=k.

- Therefore, the principle of mathematical induction holds for all positive integers.

We will employ the principle of mathematical induction in the section titled "Proving that for the Gamma function, $\Gamma(n+1) = n!$".

5.45 Proving, for the Gamma function, $\Gamma(n+1) = n!$

This will be a proof by induction (see section titled "Proving the principle of mathematical induction").

Here is the gamma function.[17]

$$\Gamma(n) = \int_0^\infty x^{n-1} e^{-x}\, dx \qquad \text{Gamma function}$$

We need to prove that for every nonnegative integer, n,

$$\Gamma(n+1) = n!$$

We can show that the assertion is true for n=0,

$$\Gamma(0+1) = 0!$$

Trivially,

$$\Gamma(0+1) = \Gamma(1) = \int_0^\infty x^{1-1} e^{-x}\, dx = \int_0^\infty e^{-x}\, dx$$

Evaluating

$$\int_0^\infty e^{-x}\, dx = -e^{-x}\Big|_0^\infty = -e^{-\infty} - (-e^{-0}) = 0 + 1 = 0! = \Gamma(0+1)$$

So, we have proven our assertion for the case for $\Gamma(1)$

Now, for the second step of our proof by induction, we'll need to show that if we know that our assertion is true for the general case of n, then we can show that it will also be true for the case of n+1.

Returning to our definition of the Gamma function,

$$\Gamma(n+1) = \int_0^\infty x^n e^{-x}\, dx$$

[17] Our proof closely follows the proof described by Adam Merberg and Steven J. Miller in their book *Approximation Methods in Statistics. Course Notes for Math 162: Mathematical Statistics.* Williams College, Williamstown, MA. August 18, 2006.

Now, we just integrate by parts.[18]

$$\Gamma(n+1) = -e^{-x}x^n\Big|_0^\infty + \int_0^\infty nx^{n-1}e^{-x}dx = 0 + n\Gamma(n)$$

So,

$$\Gamma(n+1) = n\Gamma(n)$$

We can trace the value of $\Gamma(n+1)$ all the way back to 0! by repeatedly applying the above relationship.

$$\Gamma(n+1) = n\Gamma(n) = n(n-1)\Gamma(n-1) = n(n-1)(n-2)\Gamma(n-2) = \ldots = n!$$

Thus, for all positive integers,

$$\Gamma(n+1) = n!$$

[18]See section titled "Proving integration by parts."

5.46 Proving Laplace's method

"If man had limited himself to the accumulation of facts, then science would be nothing but a sterile nomenclature and the great laws of nature would have remained unknown forever".
—Pierre Simon de Laplace

Quantum physicists commonly encounter integrals in the form: $\int_a^b e^{Mf(x)}\,dx$,. Consequently, simple methods to evaluate such integrals are welcome. The Laplace method provides the following approximation the the Laplace integral.

$$\int_a^b e^{Mf(x)}\,dx \approx \sqrt{\frac{2\pi}{M|f''(x_0)|}}e^{Mf(x_0)} \text{ as } M \to \infty.$$

Laplace's method can be easily proven using only the Taylor expansion (which we've proven) and the Gaussian integral (which we have also proven).

First, let's look at the Taylor expansion for a function, $f(x)$ around a point $f(x_0)$

$$f(x) = f(x_0) + f'(x_0)(x - x_0) + \frac{1}{2}f''(x_0)(x - x_0)^2 + \ldots$$

We'll assign x_0 as the point for which $f(x)$ is a maximum (a stationary point or constant, in this case). Therefore, the derivative of $f(x_0)$ is zero, and the second term in our Taylor expansion drops out. Since the Laplace method is an approximation method, we'll drop out everything beyond the third term, and we're left with the following approximation for our function:

$$f(x) \approx f(x_0) + \frac{1}{2}f''(x_0)(x - x_0)^2$$

Now we can substitute our approximation of $f(x)$ back into the Laplace integral:

$$\int_a^b e^{Mf(x)}\,dx \approx e^{Mf(x_0)} \int_a^b e^{+\frac{1}{2}Mf''(x_0)(x-x_0)^2}\,dx$$

Does the expression on the right look familiar? It comes very close in form to the Gaussian integral (previously proven). Let's refresh our memory.

$$\int_{-\infty}^{\infty} e^{-a(x+b)^2}\,dx = \sqrt{\frac{\pi}{a}} \qquad \text{The definite integral of a Gaussian}$$

But here's a problem. In our Taylor approximation, we see that the exponential term is a positive expression, while our Gaussian calls for a negative exponent. What can we do to rectify the situation?

As you recall, we chose x_0 as a maximum. Hence the double differential (f'') about x_0 must be a negative number (it must be declining). When we evaluate the absolute value of f'' we reverse the sign of its value. To compensate, we must reverse the sign of the term, leaving the value of the equation for $f(x)$ unchanged, but bringing the right side of the equation into a form that fits the Gaussian integral. Now, we can rewrite the Laplace equation as:

$$\int_a^b e^{Mf(x)}\,dx \approx e^{Mf(x_0)} \int_a^b e^{-\frac{1}{2}M|f''(x_0)|(x-x_0)^2}\,dx$$

And, applying the Gaussian integral formula, we get the equation for Laplace's method.

$$\int_a^b e^{Mf(x)}\,dx \approx \sqrt{\frac{2\pi}{M|f''(x_0)|}}e^{Mf(x_0)} \text{ as } M \to \infty.$$

Note that the approximation gains precision when M is large.

5.47 Proving convolution commutes

We need to prove that $f(t) \star g(t) = g(t) \star f(t)$, where $\star$ represents the convolution operation. The proof is straightforward.

The definition of convolution of two functions is,

$$f(t) \star g(t) = \int_0^t f(\tau)g(t-\tau)d\tau$$

There is nothing to stop us from making the substitution $u = t - \tau$. Doing so give us,

$$f(t) \star g(t) = \int_0^t g(u)f(t-u)du$$

The left-side of the equation is equivalent to $g(t) \star f(t)$

Therefore,

$$f(t) \star g(t) = g(t) \star f(t)$$

This proves that the convolution operator commutes. We note the Laplace transform, a closely related operator, does not commute.[19] When commutation is required, we exploit the relationship between convolution and the Laplace transform.

When,

$$\mathcal{L}\{f(t)\} = F(s) \text{ and } \mathcal{L}\{g(t)\} = G(s)$$

Then,

$$\mathcal{L}^{-1}\{F(s)G(s)\} = f(t) \star g(t) = g(t) \star f(t)$$

We see that the commuting convolution operator substitutes for the inverse of the Laplace transform operator.

[19]Let's not confuse the Laplace transform operator, $\mathcal{L}$ with the Laplacian, ∇^2.

5.48 Proving Stirling's approximation

Stirling's approximation for large numbers (N) is:

$$N! \approx \sqrt{2\pi N} N^N e^{-N}$$

This approximation, named for James Stirling, (1692-1770), is extremely useful in combinatorics, where factorials pop up everywhere. We can prove Stirling's approximation in just a few when we employ Laplace's method and the factorial representation of the Gamma function, both of which have been previously proven.[20]

Applying the Gamma function,

$$N! = \Gamma(N+1) = \int_0^\infty e^{-x} x^N \, dx.$$

Let $x = Nz$

$$dx = Ndz.$$

$$\begin{aligned} N! &= \int_0^\infty e^{-Nz} (Nz)^N N \, dz \\ &= N^{N+1} \int_0^\infty e^{-Nz} z^N \, dz \\ &= N^{N+1} \int_0^\infty e^{-Nz} e^{N \ln z} \, dz \\ &= N^{N+1} \int_0^\infty e^{N(\ln z - z)} \, dz \qquad \text{Equation 1} \end{aligned}$$

Let,

$$f(z) = \ln z - z$$

Then take the first differential of $f(z)$

$$f'(z) = \frac{1}{z} - 1,$$

[20]See sections titled "Proving Laplace's method", and "Proving, for the Gamma function, $\Gamma(n+1) = n!$".

Then take the second differential of $f(z)$,

$$f''(z) = -\frac{1}{z^2}.$$

Evaluating at the maximum value of f(z), $z = 1$, we have $f(1) = -1$, $f'(1) = 0$, $f''(1) = -1$, and $|f''(1)| = 1$

Now, let's look again at Laplace's method (see section titled "Proving Laplace's method").

$$\int_a^b e^{Mf(x)}\,dx \approx \sqrt{\frac{2\pi}{M|f''(x_0)|}}e^{Mf(x_0)} \text{ as } M \to \infty.$$

We see that the integral (right side of Equation 1, above), fits the Laplace integral. All we need do is directly apply the Laplace method formula to the integral of equation 1 at the maximal value of f(z), wherein $x_0 = 1$, $|f''(x_0)| = 1$, and $f(x_0) = -1$, to yield the Stirling approximation.

$$N! \approx N^{N+1}\sqrt{\frac{2\pi}{N}}e^{-N} = \sqrt{2\pi N}N^N e^{-N}.$$

5.49 Proving Green's theorem

Green's theorem, named for George Green (1793-1841), equates the net sum of all the micro-circulations held within a surface (i.e., the circulation within an area or a volume) to the flow around its boundary (i.e., the perimeter of a two-dimensional area or the surface of a three-dimensional volume). The circulation of F around C can be expressed as,

$$\int_C F\dot{d}s \tag{5.4}$$

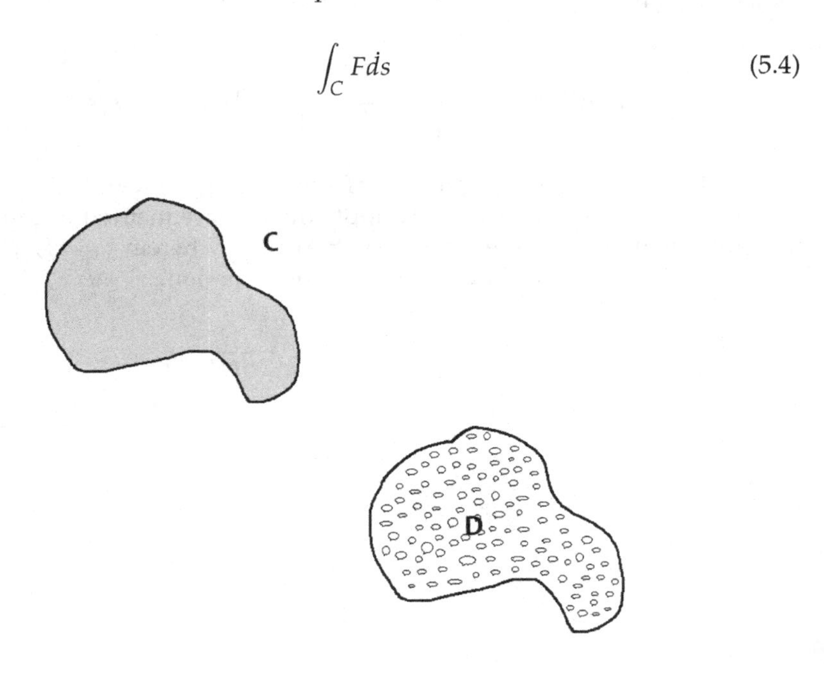

FIGURE 5.15: Green's theorem, illustrated. The flow around the perimeter of an enclosed area, C, is equal to the sum of all of the micro-flows within the area.

On a simplistic level, Green's theorem seems obvious. Let's imagine a homogenous fluid circulation that is confined snuggly within a perimeter. Now, let's ask ourselves what energy might be associated with the confined circulation. Remembering that energy is force time distance, we can infer that the energy produced by fluid circulating with a force, F, around an enclosing line, C, is just the force, F, produced by the circulating fluid over the distance around its enclosure.

Now, let's imagine that the fluid within the perimeter is composed of two different circulations. The energy derived from the circulation is equal to the sum of the forces of the contributing circulations times the sum of the distances around the perimeter. The shared border between the two micro-circulations is internalized and does not contribute to the energy expended at the outer perimeter.

We can divide the total circulation into any chosen number of micro-circulatory parts, as we wish. The results will always be the same. Namely, the net energy of circulation over the enclosing surface is $\int_C F\dot{d}s$.

We could stop here if we wished. But we should clarify what we mean when we refer to the force produced by a circulation. In the case of a circulating fluid, the forces involved have direction (i.e., can be described as vectors). When the circulating fluid is 2-Dimensional (e.g., having x and y coordinates) and confined to an area, the function that describes the circulation will have x and y parameters, but no z parameter. That is to say that the position of points in the rotating micro-circulation will be a function of x and y. The surprising and somewhat unintuitive consequence is that the force vector arising from rotation in the xy plane will point in the z-direction.

To get an intuitive grasp of the forces that arise from an area of circulation, think of an area on the surface of water that is moving in a circle around a central point. The circulating water will produce a downward vortex or whirlpool, in the direction perpendicular to the surface motion. Likewise, when a tornado passes over land, the direction of the wind will be parallel to the ground, but objects in the tornado's path will be lifted upward into its funnel. When a tornado passes over water, water will be lifted into the funnel of the tornado, producing a waterspout. In every case, the force of the rotating fluid (air, in the case of a tornado) points in the direction perpendicular to rotation.

A derecho is a wall of moving air, sometimes traveling with tremendous speed, that can travel over land, with devastating effect. Flattened trees, bushes, and grasses lie in the wake of the derecho. The affected vegetation will lie straight and flat, all oriented in the same direction. In a derecho, debris is not hurtled high into the sky, because there is no vertical lift associated with a flat wall of air. The unidirectionality of the fallen debris, associated with a non-rotational force, is totally unlike the residue pattern left by a tornado. A tornado will lift cars and trees into the sky and debris will lay scattered, with a radial twist.

In summary, the force produced by a radial circulation (or microcirculation) is perpendicular to the direction of the circulation. To determine the magnitude of the force produced by a circulating fluid, we turn to a vector operator known as the curl symbolized as "$\nabla \times$". A curl converts a vector field into another vector field with each point in the vector field representing rotation at the point. In matrix form, the curl of a function, F, with x,y, and z parameters is:

$$\nabla \times \mathbf{F} = \begin{vmatrix} \hat{i} & \hat{j} & \hat{k} \\ \frac{\partial}{\partial x} & \frac{\partial}{\partial y} & \frac{\partial}{\partial z} \\ F_x & F_y & F_z \end{vmatrix} \tag{5.5}$$

Here, $\hat{i}$, $\hat{j}$, and $\hat{k}$ are the unit vectors in the x, y, and z directions respectively The value of the curl can be computed by the following formula.

$$\nabla \times \mathbf{F} = \left(\frac{\partial F_z}{\partial y} - \frac{\partial F_y}{\partial z}\right)\hat{\mathbf{i}} + \left(\frac{\partial F_x}{\partial z} - \frac{\partial F_z}{\partial x}\right)\hat{\mathbf{j}} + \left(\frac{\partial F_y}{\partial x} - \frac{\partial F_x}{\partial y}\right)\hat{\mathbf{k}} \tag{5.6}$$

In the case where the circulation of a small area is confined to the xy plane (i.e., an area delimited by dx and dy), the vectors of in the x and y direction evaluate to zero, and the vector of the curl is entirely in the z direction (the $\hat{k}$ unit component).

$$\nabla \times \mathbf{F} = \left(\frac{\partial F_y}{\partial x} - \frac{\partial F_x}{\partial y}\right)\hat{k} \tag{5.7}$$

Green's theorem evaluates to,

$$\int_C F ds = \iint_D \nabla \times \mathbf{F} = \iint_D \left(\frac{\partial F_y}{\partial x} - \frac{\partial F_x}{\partial y}\right)\hat{k} \tag{5.8}$$

Here, "D" is the area enclosed by the perimeter, "C".[21]

[21]Stokes' theorem (named for George Stokes) takes Green's theorem (which applies to areas), and extends it generally to closed paths. Stokes' theorem is commonly applied to three dimensional enclosures.

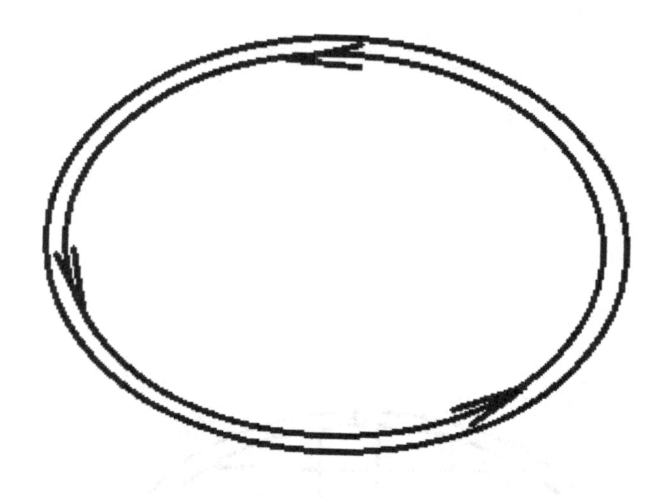

FIGURE 5.16: A perimeter within which there is one area of circulating fluid. The energy derived from the circulation is just the force produced by circulation times the distance around the perimeter.

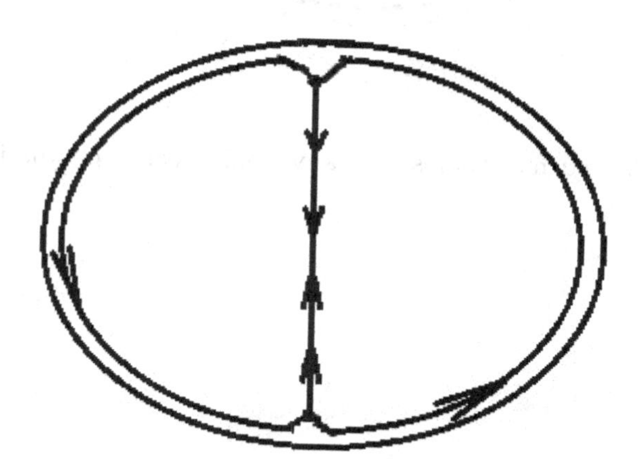

FIGURE 5.17: A surface enclosing two areas of circulating fluid.

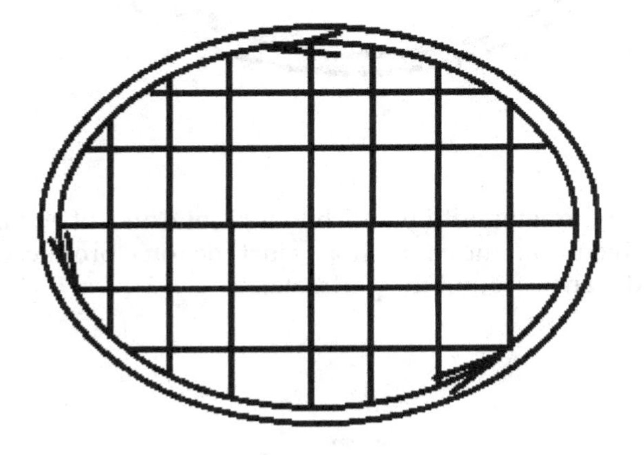

FIGURE 5.18: A surface enclosing many small areas of circulating fluid.

5.50 Proving Fourier transforms of Gaussians are Gaussians

We previously discussed the Gaussian integral (see section titled "Proving that the Gaussian integral = $\sqrt{\pi}$".) The Gaussian integral is represented by a bell-shaped curve and is represented by the following equation:

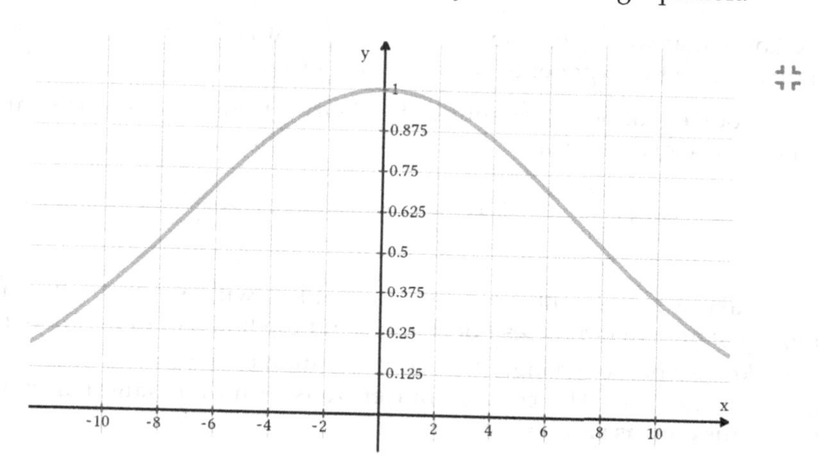

FIGURE 5.19: The familiar bell-shaped curve of the Gaussian distribution.

$$\int_{-\infty}^{\infty} e^{-x^2}\,dx \qquad \text{The simplest Gaussian integral}$$

The Gaussian function that is being integrated is, of course, e^{-x^2}. Let's evaluate the Fourier transform of the simplest Gaussian function.

$$\begin{aligned}\hat{f}(\omega) &= \int_{-\infty}^{\infty} f(x)e^{-ix\omega}\,dx \\ &= \int_{-\infty}^{\infty} e^{-i\omega x}e^{x^2}dx\end{aligned}$$

Now, let's multiply the right side of the equation by $e^{\omega^2/4} \div e^{\omega^2/4} = 1$. Because we're dealing with a constant, we can put the bottom half of the fraction inside the integral and we can put the top half of the constant outside the integral, recognizing that $\frac{1}{e^{\omega^2/4}} = e^{-\omega^2/4}$.

$$= \int_{-\infty}^{\infty} e^{-i\omega x} e^{x^2} dx$$
$$= e^{\omega^2/4} \int_{-\infty}^{\infty} e^{-i\omega x} e^{x^2} e^{-\omega^2/4} dx$$

But we know that we can equate $e^{-i\omega x} e^{x^2} e^{-\omega^2/4}$ with $e^{-(x^2+i\omega x+(i\omega/2)^2)}$, and the expression in the exponential can be easily factored.

Therefore, our equation representing the Fourier transform of our simple Gaussian can be expressed as:

$$\hat{f}(\omega) = e^{\omega^2/4} \int_{-\infty}^{\infty} e^{-(x+i\omega/2)^2} dx$$

We can make a simple substitution of variables, with x substituting for $x + i\omega/2$. We note that dx does not change, and the limits of integration, being $\pm\infty$ likewise do not change. Now, our equation becomes $e^{\omega^2/4}$ times the integral of a Gaussian. **Hence, the Fourier transform of a Gaussian yields another Gaussian, as shown:**

$$\hat{f}(\omega) = e^{\omega^2/4} \int_{-\infty}^{\infty} e^{-x^2} dx$$

But we have previously shown that the integral of the value of the integral of a Gaussian is $\sqrt{\pi}$. So, the Fourier transform of our original Gaussian is:

$$\hat{f}(\omega) = e^{\omega^2/4} \sqrt{\pi}$$

The normal curve of statistics, often referred to as the Poisson distribution (named for Siméon Poisson), is a type of Gaussian curve represented by the following function:

$$f(x) = e^{-\frac{1}{2}(\frac{x}{\sigma_x})^2} \qquad \text{The Poisson distribution}$$

When we take the Fourier transform of the Poisson distribution, we get:

$$\hat{f}(k)$$

$$= \frac{1}{2\pi}\int_{-\infty}^{\infty} e^{ikx} g(x)dx$$

$$= \frac{\sigma_x}{\sqrt{2\pi}} e^{-\frac{1}{2}\sigma_x^2 k^2}$$

Now, let's substitute $\frac{1}{\sigma_k}$ for σ_x

$$= \frac{1}{\sqrt{2\pi}\sigma_k} e^{-\frac{1}{2}(\frac{k}{\sigma_k})^2}$$

We have just shown that the Fourier transform of a Poison distribution produces another Poisson distribution, but **the standard deviations of the original Poisson distribution and its Fourier transform are reciprocals of one another.** Somewhat surprisingly, we have proven one of the most significant relationships in physics. That is to say, for statistical representations of Fourier conjugates, the standard deviation of one conjugate is the reciprocal of the standard deviation of the other conjugate.

Let's explore the consequences of our findings:

- When a Gaussian function is a narrow bell curve, then the Fourier of the Gaussian function (its' Fourier conjugate, also a Gaussian) will be represented by a wide bell curve.
- When a Gaussian function is a wide bell curve, its Fourier transformation will be a narrow bell curve.
- When the Gaussian variable is measured in time, the Fourier of the Gaussian variable will be measured in frequency.
- When a Poisson Gaussian is measured in frequency (including Gaussians that are themselves the result of a Fourier transformation), its conjugate will be a Gaussian that represents the distribution of time.
- The Fourier of a short impulse (in time) will yield a Gaussian that is spread out widely over frequencies.
- The Fourier transform of a narrow frequency Gaussian will yield a Gaussian that is spread out over time.
- The same relationship applies to waves, wave packets, signals, and the Fourier conjugates that represent relationships described by the Heisenberg uncertainty principle, between energy and momentum or position and time.

In the case of the narrowest possible Gaussian, the Dirac delta function (a straight vertical line), its Fourier transform is the widest possible distribution of frequencies, a straight horizontal line in the frequency domain, representing all frequencies.

"Last words are for fools who haven't said enough".
—Last words of Karl Marx

Index

Printed in the United States
by Baker & Taylor Publisher Services